Oceans: A Very Short Introduction

VERY SHORT INTRODUCTIONS are for anyone wanting a stimulating and accessible way into a new subject. They are written by experts, and have been translated into more than 45 different languages.

The series began in 1995, and now covers a wide variety of topics in every discipline. The VSI library now contains over 500 volumes—a Very Short Introduction to everything from Psychology and Philosophy of Science to American History and Relativity—and continues to grow in every subject area.

Very Short Introductions available now:

For more information visit our website

www.oup.com/vsi/

Dorrik Stow

OCEANS

A Very Short Introduction

OXFORD
UNIVERSITY PRESS

OXFORD
UNIVERSITY PRESS

Great Clarendon Street, Oxford, OX2 6DP,
United Kingdom

Oxford University Press is a department of the University of Oxford.
It furthers the University's objective of excellence in research, scholarship,
and education by publishing worldwide. Oxford is a registered trade mark of
Oxford University Press in the UK and in certain other countries

© Dorrik Stow 2017

The moral rights of the author have been asserted

First edition published in 2017

Published in the United States of America by Oxford University Press
198 Madison Avenue, New York, NY 10016, United States of America

British Library Cataloguing in Publication Data
Data available

Library of Congress Control Number: 2017936754

ISBN 978-0-19-965507-6

Printed and bound by
CPI Group (UK) Ltd, Croydon, CR0 4YY

For Terry

Contents

Oceans

Preface

As I write the final part of this manuscript, I am enjoying a crisp white wine under the late afternoon sunshine in Malta and gazing out across the blue Mediterranean. It is particularly apposite to reflect on one of the people whose work has inspired my own interest in the oceans—a Maltese diplomat, scholar, and university professor called Arvid Pardo. From 1964 to 1971, he was the first Permanent Representative of Malta to the United Nations, where he became known as the 'Father of the Law of the Sea Conference'. In November 1967, he made an inspirational speech to the General Assembly in which he called for international regulations to ensure peace at sea, to prevent further pollution, and to protect ocean resources. He proposed that the seabed beyond national jurisdiction (then only 3 nautical miles from the coastline) should constitute part of the common heritage of mankind and that some of the ocean's wealth be used for the direct benefit of poorer nations. As of June 2016, 167 countries and the European Union have joined the Law of the Sea Convention.

The oceans cover 71 per cent of the planet and our use and abuse of this ocean world is gathering momentum. The first transatlantic telegraph cable was laid as early as 1858, and now parts of the oceans are veritably crisscrossed with communication cables, as well as oil and gas pipelines and other subsurface installations. Today, 99 per cent of all internet data is transmitted by submarine cables that stretch over hundreds of thousands of kilometres in length and reach as deep as Everest is high. Oil exploration took its first tentative steps into the shallow offshore realm in 1924 and has since moved into progressively deeper waters. Over 50 per cent of oil and gas discoveries in the last decade have been made in the deep oceans, and companies are routinely drilling on the continental slope in water depths over 2500 metres. The 21st century is witnessing an explosive increase in renewable energy production, equalling the growth trajectory of oil and gas in the last century. Offshore wind turbines, tidal barrages, and energy from waves and other marine sources are an important part of that growth.

Food from the oceans has preoccupied coastal communities from the very dawn of civilization. For more than a billion people, fish and shellfish are still the dominant part of their essential protein consumption (over 90 per cent for some). For the world as a whole this figure is around 10–20 per cent. This is a vast and vital commodity that will become even more significant as a burgeoning world population demands ever more food resources, when massive fish factory fleets ply the high seas, and when the touchpaper to the blue revolution in mariculture is set to be lit. The oceans also offer an almost infinite source of water—from which clean drinking water can be extracted through desalination. A safe and sufficient water resource for everyone is perhaps the greatest challenge this century.

The mineral riches of the ocean are vast and barely tapped. Fifty years ago the mining industry was clamouring at the gateway to the deep ocean—manganese nodules had been discovered in great abundance on the deep-sea floor, comprising a wide range of strategic metals. But still no inventory exists of these nor of the more concentrated polymetallic chimneys and edifices recently discovered around hot-water vents. Perhaps even more significant is the abundance of rare-earth elements now believed to be concentrated in deep-sea clays. These are an unusual group of rare metals—yttrium and the lanthanide series—that are being consumed in ever-increasing amounts for everyday products such as computer memory chips, mobile phones, batteries, DVDs, and much more. Terrestrial resources are very limited for all these minerals.

The seas are the second oldest battlefield on earth. Though often glorified in history books, I firmly hope and believe that they will ultimately be *confined* to those books and even to legend. Although the sea today still supports nearly 10,000 naval vessels, this number is less than 10 per cent of the 100,000 ships worldwide. More than half of these are merchant ships for peaceful trade between nations. The world's largest and still one of the fastest growing multinational industries is tourism. Coastal or island destinations and maritime leisure activities, passenger ferries, and luxury cruises, together account for a huge proportion of global tourist trade. Oceans provide a receptacle for our human waste, a capacity to reabsorb our carbon dioxide emissions from the atmosphere, a cleanser of both sewage and oil spills.

It is vital that we also recognize our ability as a species to alter the Earth's environment, understand just how we are doing so, and accurately predict the likely repercussion of our actions. The oceans are the single biggest driver and buffer of global warming. They also hold the most complete record of past climate changes that have taken place and of their effects. The temperature of the sea helps control everyday weather patterns, generates hurricanes and tropical cyclones, and leads to regional droughts or excessive monsoonal rainfall. The oceans are under serious threat of near irreversible pollution in certain coastal areas and marginal or enclosed seas. Everywhere, the oceans are becoming more acidic and this chemical change profoundly affects marine biota—for example, coral bleaching, coral death, and the decline of the great kelp forests.

Whatever the societal challenge for the oceans—communications, energy, food, water, minerals, trade, tourism, waste disposal, natural hazards, global warming, or environmental degradation—we need to understand our ocean world better. This is the oceanographer's challenge as she or he strives to address the major scientific questions of the day and for the future. It has been my personal challenge for the past forty years.

Scientific understanding of the oceans is in its infancy but we are learning fast. This book aims to set out what we *do* know about this watery world: how oceans originate, evolve, and change; the shape of the seafloor and nature of its cover; the physical processes that stir the waters and mix such a rich chemical broth; the inseparable link between oceans and climate; and the complex web of marine life. It will attempt to provide an easily comprehensible insight into a complex world, emphasizing the very important link between scientific knowledge of the oceans and human exploitation of their ultimately limited resources and fragile environment.

Acknowledgements

First and foremost, my eternal thanks to Claire for her love and support, and to my children—Jay, Lani, and Kiah. I am also very grateful to Latha Menon, Jenny Nugee, and the whole team at Oxford University Press for their advice, encouragement, and patience throughout this project. I would especially like to thank Latha Menon and Murray Roberts for their careful reading and helpful comments on the manuscript. My great appreciation goes to the many colleagues and institutes around the world who have shared their wisdom with me. Equally, I have had the privilege to participate in many scientific expeditions on land and at sea, which have greatly improved my knowledge of the oceans. Heriot-Watt University has provided overall support during the writing of this book.

List of illustrations

Oceans

List of tables

Chapter 1
Ocean frontier

From the dawn of history the ocean has provided a source of food, refreshed our bodies, and cleansed waste. Once a barrier to human migration it has since become a highway for transport and trade, as well as a battlefield for the long pageant of human conflict. It offers adventure and challenge to explorers and scientists, a source of philosophical and poetic wonderment, and a playground for our children. Most importantly, the oceans offer both the challenges and solutions for many of the key issues that face the world today—sea-level rise and climate change, new sources of food and water for everyone, natural resources and energy for growth and development, a new source of medicines, and safe waste recycling. Excitingly, it is still a great unknown frontier with an endless capacity to both surprise and alarm.

Grand ideas

For 5000 years and more, since cuneiform writing was first invented and records kept, philosophers and scientists have been drawn to the ocean. The Sumerians plied the waters of the Tigris and Euphrates rivers as well as the Persian Gulf and, as far as we know, were the first to use sail boats. As this cradle of civilization lay between two major rivers it is little wonder that their writings record the advent of huge floods and the resultant devastating effects on people and animals. The story of the great flood as told in the Bible some 2000 years later almost certainly originated from these Sumerian records and stories.

Among seafaring people, in particular, myths and legends concerning the sea became one way of coping with what they could not explain. The Egyptians, Greeks, Polynesians, Vikings, and countless others have all invoked powerful deities as masters of this watery world. The ocean was alternately savage and powerful, serene and beautiful—and it was the gods that made it thus.

Aristotle (384–322 BCE) was one of the first natural philosophers to take a scientific interest in the sea and its coastline, how deltas form and the 'filling' of the Mediterranean Sea. He wrote the first known treatise on marine biology. His recognition that the changes of the Earth and oceans are constant but inordinately slow in comparison to human life was especially far-sighted. He wrote: 'the distribution of land and sea in particular regions does not endure throughout all time, but it becomes sea in those parts where it was land, and again it becomes land where it was sea, and there is reason for thinking that these changes take place according to a certain system, and within a certain period'. I find this insight truly amazing for his time.

But the systematic science of oceanography really began only 250 years ago. Since then, it is from ocean science that some of the most momentous discoveries of recent times have derived. These have led to four seminal ideas that help shape our thinking today:

- The living world is one of constant evolution, extinction, and change. Life on Earth had its origins in the ocean over 3.5 billion years ago. Countless organisms have lived and perished since that time, most at the unicellular scale, but evolution slowly progressed to yield the kaleidoscopic profusion of today.

- The Earth is in continual motion, but on a geological timescale imperceptible to the human frame. Oceans are created and destroyed, continents move, mountains rise from the sea and are then returned as dust to the ocean floor.

- Changes are cyclic in nature, although the timescales vary and the past is never exactly repeated. This is true for oceans and continents, extinction and evolution, sea level and climate.

- The natural environment of ocean, earth, and atmosphere is a single interconnected entity. It is also the single most important variable for the survival of our own and other species. In the past fifty years we have come to know that we are effecting major changes on the environment, but we do not yet know how to control those changes for good.

These fundamental concepts—evolution, geological time, cyclicity, and environmental change—not only act as profound paradigms in natural science, but have also reached into popular consciousness. They have an everyday reality that helps us realize our place in the world and to recognize the imperative for us to care for the environment and act as stewards for its limited resources.

Early exploration

Long before written records, we know that an early hominid, *Homo erectus*, had mastered crossing the relatively small stretches of water that separated one island from another along the Indonesian archipelago and established a presence on the island of Flores. There is then a huge gap in marine archaeological knowledge before the first waves of *Homo sapiens* migration out of Africa around 70,000 years ago, in part by island-hopping across the southern Red Sea into what is now the Middle East. By about 50,000 years ago they had ventured further from the Indonesian archipelago to colonize Australia.

The most remarkable story of early exploration by sea began some 5000 years ago with the spread of the Polynesian people from South East Asia, ultimately colonizing 10,000 islands across

26 million square kilometres of the Pacific Ocean. They were master sailors who reached the world's remotest islands, including the Hawaiian chain and Easter Island. They navigated by the stars and Moon, the waves and winds, the clustering of marine life and the flight of birds, and passed all this knowledge down from generation to generation by word of mouth.

At about the same time, the Egyptians were building ships for river and coastal trade, while the Phoenicians became true maritime masters of the Mediterranean region, exploring and trading to every corner of their realm, and even as far as England to the north and around Africa to the Indian Ocean. On the other side of the world, the Ciboney Indians were the first to colonize the Caribbean islands from their South American homeland, soon to be replaced by the Arawaks from the Orinoco Delta region. At this time too, the Chinese were plying their trade, language, and culture through South East Asia, and by 400 BCE had invented the magnetic compass (Figure 1).

Voyages of discovery

Exploration and discovery by sea has grown ever stronger through the last two millennia, and has often been glamorized as expeditions inspired by a sense of adventure or to advance knowledge. But the truth was far more mercenary and military. It was driven by the search for and opening up of new and lucrative trade routes around the world and then protecting them. It was accompanied by piracy and pillage, and soon led to unrivalled exploitation of the lands 'discovered'. That ocean science advanced during this time was merely a fortuitous by-product.

Sailing from Scandinavia in their characteristic longships, the Vikings were feared warriors in the North Atlantic. They landed suddenly, raided at will, and returned swiftly to sea. They were also fine sailors, who had colonized Iceland by 700 CE, Greenland soon after, and Newfoundland in North America by 1000 CE. They

2000	AGE of SCIENCE & TECHNOLOGY	Circa 2000: focus on climate, resources, oceans Circa 1980s: new environmental awareness Circa 1970s: drilling for deepwater oil and gas 1968: International Scientific Drilling begins DSDP *Glomar Challenger* launched
1900		
1800	OCEAN SCIENCE	1850–1950 Many national scientific ocean expeditions and centres established 1872–79 HMS *Challenger* expedition 1831–36 HMS *Beagle* expedition (with Darwin)
1700		1768–79 James Cook's voyages (esp. Pacific) Circa 1750: onset of systematic ocean science
1600	EUROPEAN "AGE of DISCOVERY"	Many European expeditions for trade, wealth creation, and empire building across the world Global circum-navigation, charting the oceans Incidental ocean science
1500		
1400	CHINESE TRADING	Chinese fleets trade across SE Asia and the Indian Ocean
1000 CE	VIKING VOYAGES	Expeditions to colonise Iceland, Greenland and parts of North America
0	GREEK INFLUENCE	Greek dominance in the Mediterranean First(?) scientific interest in the oceans
1000	CHINESE	Invention of magnetic compass
2000	PHOENICIANS	Mediterranean masters, circum-navigate Africa
3000	POLYNESIANS	Master Sailors – extensive colonisation of widely dispersed islands across the Pacific Ocean
	CIBONEY INDIANS	Colonise Caribbean Islands
4000	SUMERIANS and EGYPTIANS	Coastal trade and fishing, first sail boats
BCE	PRE-HISTORY	70,000 BP Homo sapiens expands out of Africa Reaches Australia by 40,000 BP

Ocean frontier

1. **Ocean exploration timeline.**

explored the Mediterranean and Black Seas and even sailed around the southern tip of Africa. Arab sailors were masters of the Indian Ocean at this time, supplementing their overland trade routes by marine expeditions. They developed a good understanding of the monsoon climate and its effects on conditions at sea, and borrowed the magnetic compass from the Far East.

The Chinese already had a long seafaring tradition. Chinese expansion reached a zenith during the Ming Dynasty, in the earliest part of the 15th century, when Admiral Zheng He commanded a magnificent fleet of 317 ships and 37,000 men, built to carry the finest materials and objects to distant lands. The intention appears to have been to give away such treasures and so demonstrate the wealth and superiority of Chinese civilization. Many of the technical advances they had developed—such as a central rudder, watertight compartments, and sophisticated sails that could be operated from the deck—filtered through into European designs.

European exploration began rather later. By the early 15th century, explorers and traders had reached China, India, and North Africa, although still no one from the Old World had any idea that America, Australia, Antarctica, or Greenland existed. There then began a century of expansion and exploration, of amazing sea voyages to unknown worlds, of new ideas and inventions. This was all part of the Renaissance period in Europe and, with a peculiarly Eurocentric view of the world, has become known as the *Age of Discovery*.

Prince Henry of Portugal was an early visionary who saw that ocean exploration could open new trade routes and yield great wealth in the process. He established a centre at Sagres for the study of marine science and funded many Portuguese expeditions to the South Atlantic. All captains were required to compile detailed charts wherever they sailed in good weather or poor, and so when a storm blew the Portuguese explorer Bartolomeu Dias around the southern tip of Africa in 1487 this route was finally rediscovered. The following year Vasco de Gama reached India.

Christopher Columbus, an Italian master mariner, attempted to reach the East by sailing west around the world. Having a misconceived view of the Earth as only about half its actual size, he mistook his first landfall on a small Caribbean island off Central America for his goal of India or Japan. It was here that he met descendants of the Arawak Indians who had colonized the region at least 5000 years before. Other explorers quickly followed. John Cabot sailed from England to Newfoundland in 1497, also seeking a new route to the East. In 1499, another Italian, Amerigo Vespucci, reached the South American coast, sailing as far south as the Amazon River—America was soon to be named after him.

In 1519, Ferdinand Magellan set off from Portugal with a flotilla of five rather ageing ships and 230 experienced seamen. Although Magellan himself was killed on the small Philippine island of Mactan, one of his captains, Sebastian del Cano, struggled on, eventually reaching Seville in September 1522—with only one ship and eighteen men remaining. Completion of this long and extremely arduous voyage was undoubtedly a crowning achievement of the time and probably the first circumnavigation of the world. Unfortunately its accomplishment marks the beginning of a far less pleasant period of global history—a time when the European powers set out to exploit the lands they had newly discovered.

Pioneers of ocean science

From the observations and insights of Aristotle to the detailed charts commissioned by Prince Henry of Portugal, from the first-hand knowledge of oceans by those who had carefully constructed maps that became ever more accurate with time, we gained a wealth of *incidental* ocean science. But by the early 18th century there was a reawakened thirst for scientific discovery, and we see the birth of true scientific oceanography. Luigi Marsigli compiled his

Histoire Physique de la Mer, the first book to deal exclusively with the ocean. Leonard Euler published his important mathematical understanding of the lunar control on tidal movement, and Benjamin Franklin charted the Gulf Stream for the first time.

The systematic acquisition of scientific data at sea was pioneered during three long and remarkable voyages led by Captain James Cook from Plymouth, England, between 1768 and 1780. For two of these, he had the added advantage of the most sophisticated and accurate chronometer of the age, specially designed by clockmaker John Harrison in order to reliably determine longitude at sea (latitude was already easy to calculate from the stars). Cook's navigation was therefore outstanding, so that the carefully gathered data on geography, geology, bathymetry, marine biota, tides, currents, water temperature, and salinity, from as far south as 71°S where he first encountered icebergs and pack ice that surrounded Antarctica, were all the more valuable for those who tried to bring them into some scientific sense of order.

Many more oceanographic expeditions followed through the 19th-century period of European empire-building, two of which were especially significant. The voyage of the survey ship HMS *Beagle* between 1831 and 1836, with a mission to map the South American coastline, was joined by the young Charles Darwin as ship's naturalist. He explained how volcanic islands and coral atolls formed in the midst of oceans, and did much important work on marine organisms, especially barnacle biology. But it was his very detailed observations of subtle variations in both fossil and living species that ultimately led to his seminal work *On the Origin of Species*, published in 1859.

The second voyage of immense significance, and for which the term 'oceanography' was first coined to describe its mission, was the round-the-world voyage of HMS *Challenger* between 1872 and 1876 (Figure 2). This expedition was a purely scientific mission and an unqualified success. It covered 127,600 kilometres,

2. HMS *Challenger*.

took thousands of depth soundings, seafloor samples of sediment and biota, and measurements of water temperature, salinity, and density. The *Challenger Report* of the expedition's findings compiled by Sir John Murray amounted to fifty volumes. It founded and forced the pace of oceanography as a science, established the enormous economic and practical significance of such work, demonstrated the need for dedicated multidisciplinary missions, and fostered international cooperation.

Global challenge

Oceanographic research today is as exciting and challenging as ever. As the pace of discovery accelerated through the 20th century, and as new technologies were applied to ocean science, so the body of knowledge about the seas grew exponentially. But this has simply opened windows to the great unknown, providing a tiny vista on a new frontier of exploration. We are now so much better placed to know the scientific questions to ask, and so much better prepared, with a whole raft of technology at our disposal, that adventuring into the ocean realm can be immensely rewarding.

Oceanography is *big science*, necessarily so because of the size of the canvas on which it is undertaken, the nature of the equipment and sea-going vessels required, and the scale of the questions being tackled. It has always needed a cooperative rather than individual spirit, with the backing of large institutions or governments. One of the biggest scientific investigations ever conducted began in 1968 as the Deep Sea Drilling Project and still continues today, nearly five decades later, as the Integrated Ocean Drilling Program. Since the *Glomar Challenger* scientific drillship set sail in 1968, the *JOIDES Resolution*, *Chikkyu*, and others have followed (Figure 3). Over 2000 holes have been drilled deep below the seafloor, nearly 2 million metres of sediment core have been recovered, and some of the most fundamental questions about planet Earth have been posed and answered.

Oceanography is also a very *practical science*, with very real application to the needs of society. Whatever the societal challenge

3. *JOIDES Resolution* scientific drillship.

for the oceans—communications, energy, food, water, minerals, trade, tourism, waste disposal, natural hazards, global warming, or environmental degradation—we need to understand better our ocean world.

This is the oceanographer's personal challenge as she or he strives to address the major scientific questions of the day and for the future. What moulds the shape of the ocean floor and its interaction with the seawater above? What is the source and flux of sediment to and within the oceans? Can we read the ocean's long-term memory as a barometer of environmental change? How are the oceans stirred and on what timescale? How does the ocean maintain its chemical balance and what might upset this ability? Can we understand and model the ocean–climate nexus and can this help mitigate global warming? What new species of organism and momentous discoveries in the field of biology will deep-sea exploration next reveal? There are many more questions, of course. None of them is easy to answer—but the rewards are of scientific, economic, and societal importance.

The task of addressing these questions falls to research oceanographers and their students at universities and institutes across the world. Typically, they work together in interdisciplinary teams, often involving cooperation between several institutes and different countries, and participation in large-scale regional or global projects. Oceanography is still very much a field science on a huge canvas (Figure 4, Table 1). The practical difficulties involved are easily matched by the personal hardships endured—cold, damp, nausea, and frustration. It can be a remote and lonely science, facing the wild open ocean or the danger of deep-sea diving. But the thrill of discovery, the knowledge of its significance, and the undeniable beauty of the ocean world are ample reward for engagement in the ocean quest.

Oceans

4. Global oceans map.

Table 1 Ocean data

Atlantic Ocean

Area	82,000,000 km^2
Average depth	3300 m
Maximum depth	8605 (Milwaukee Deep)
Volume	321,930,000 km^3
Oldest ocean crust	175 million years (Middle Jurassic)
Marginal seas	Gulf of Mexico, Caribbean, North Sea, Mediterranean Sea
	Labrador Sea, Baltic Sea

Indian Ocean

Area	73,600,000 km^2
Average depth	3890 m
Maximum depth	7450 m (Java Trench)
Volume	292,131,000 km^3
Oldest ocean crust	140 million years (Early Cretaceous)
Marginal seas	Red Sea, Gulf of Aden, Persian Gulf

Pacific Ocean

Area	166,000,000 km^2
Average depth	4280 m
Maximum depth	10,924 m (Challenger Deep)
Volume	723,700,000 km^3
Oldest ocean crust	165 million years (Middle Jurassic)
Marginal seas	South China Sea, East China Sea, Java Sea, Celebes Sea, Banda Sea, Sea of Japan, Sea of Okhotsk, Bering Sea, Coral Sea, Tasman Sea

Arctic Ocean

Area	12,173,000 km^2
Average depth	990 m
Maximum depth	5608 m (Molloy Deep)
Volume	17,750,000 km^3
Oldest ocean crust	55 million years (Paleogene)
Marginal seas	Beaufort Sea, Chukchi Sea, East Siberian Sea, Laptev Sea, Kara Sea

(continued)

Table 1 Continued

Southern Ocean

Area	35,000,000 km^2
Average depth	3350 m
Maximum depth	7235 m (South Sandwich Trench)
Volume	117,250,000 km^3
Oldest ocean crust	140 million years (Early Cretaceous)
Marginal seas	Weddell Sea, Bellingshausen Sea, Ross Sea

Chapter 2
Ocean transience

Once formed from the primitive Earth environment, the oceans as well as the continents have been constantly moving, changing, and evolving, at a rate perceptible only in geological time. Oceans are defined as much by the shape and features of the underlying seafloor (the ocean basin made of ocean crust) and their cover of sediment and organisms, as by the water they contain. Ocean basins are born, expand, and contract, islands rise from the sea, continents grow and divide. The land-ocean crust as well as the Earth's interior is acting as one unified system in this slow dance of time. This discovery in the middle of the last century led to the most recent great paradigm of the earth and ocean sciences—that of plate tectonics. This chapter explains the origin and evolution of oceans within the plate tectonic paradigm.

As a student at Cambridge University, I well remember a buzz in the lecture theatre as some of the scientists involved in developing the plate tectonic theory in the early 1970s (Drum Matthews, Fred Vine, Dan Mackenzie) outlined their newest results and ideas. We felt as though we were surfing on the crest of a new wave of discovery in our science—as indeed we were. Many years later, as chief scientist of the Integrated Ocean Drilling Program expedition (IODP 339) I took the *JOIDES Resolution* drillship to a region of the Atlantic Ocean between the Azores and Gibraltar, and witnessed the same tangible excitement as we realized that our drilling results had revealed something entirely new about the plate tectonic mechanism. We had recovered a weak but distinct signal of what I call the 'Earth's heartbeat'. Our records showed the near-surface response to powerful tectonic forces that were acting deep within the Earth's mantle at one-million-year intervals. We are still working on this discovery.

Origin of the oceans and atmosphere

The Earth condensed from a cloud of dust and gas around 4.5 billion years ago. As the third planet out from the Sun, it was neither too close and hot so that the surface was covered in a cocktail of poisonous vapours, nor too distant and cold so that only barren, frozen wastes prevailed. In the process of condensation and the associated nuclear inferno that ensued, the principal chemical elements that make up our planet were formed. Segregation of the heavier elements towards the centre led to a layered Earth with core, mantle, and crust, each having a distinct chemical composition (Figure 5).

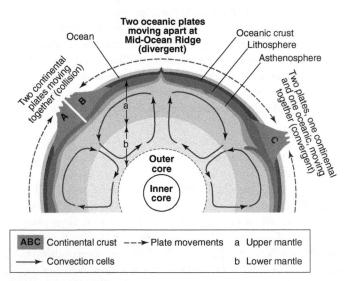

5. **Structure of the Earth.**

It is probable that the lightest elements (H, He, N, C, O) could not be retained as an atmosphere at this stage and simply boiled off into space.

The next crucial phase early in the development of the planet was differentiation of its outermost layer into oceanic and continental crust. The rock composition of these and their mode of formation are very different. Oceanic crust, the more primeval, is made up of dark volcanic rocks, now mainly basalts, which extruded at the surface from a molten interior under repeated violent eruptions. Continental crust, made up of still lighter materials—granites, metamorphic rocks, and sediments—took much longer to develop. Repeated melting and cooling allowed these progressively lighter materials to separate gradually, coagulating into the nuclei of primitive continents. Aggressive weathering by torrential rains from an early atmosphere led to the break-up of these rocks and their redistribution as sediments along the margins of the first small continents. Countless cycles repeated and slowly the continents formed, relatively more elevated due to their lower density than the surrounding proto-oceanic basins. About 7 per cent of the present-day continental crust consists of remnants of these early continents, all over 2.5 million and some as much as 3.8 million years old.

There are two competing theories for the genesis of the oceans. The first proposes that oceans formed from the gases that issued during the first phase of crustal differentiation, by a process known as de-gassing. These would have contributed to a primordial atmosphere and then, as the Earth cooled, water vapour would have condensed and fallen as rain. Gases emanating from volcanoes today include the same gases that had previously been lost into space, including water vapour, carbon dioxide, hydrogen, nitrogen, sulphur dioxide, and several others. These are released by the melting of the rocks and minerals in which they are bound. The second theory is that most of the water fell from space as a component of ice-laden comets during an early period of intense bombardment. Today, around 30,000 tonnes of water falls to Earth every year in a fine rain of cometary particles. The gradual change in composition of the oceans from largely freshwater to the saline brines we know today is a complex story of the interaction between the Earth's rocks, hydrosphere, atmosphere, and biosphere.

Jigsaw puzzle of the plates

The plate tectonic concept is elegant and simple, like many of the grandiose theories that stand the test of time. It holds that the outer layer of the Earth, between 120 and 180 kilometres thick, is made up of a series of very large and some smaller, rigid plates that are in constant motion with respect to one another (Figure 6). These are known as *lithospheric plates* (from the Greek *lithos* meaning stone). These plates ride on a weaker, hotter, almost molten (or partially molten) layer known as the *asthenosphere* (from Greek *asthenes* meaning weak). The asthenosphere, like the lithosphere, is around 100–200 kilometres thick.

Not only do plates move, but their size and shape constantly changes as new material is continually being added and old material consumed. This takes place at plate boundaries, irregular and interlocking cracks and healed sutures that scar Earth's outer crust, and along which the majority of tectonic activity occurs—earthquakes and volcanoes, hot springs and heat loss from the Earth's interior. At divergent plate boundaries (spreading centres) new crust is being added, while at convergent margins older crust is consumed or deformed. At transform boundaries, the gigantic, rigid plates scrape and grind past one another with enormous force.

Plates move about the outer Earth driven in large part by convection. This is a process, most familiar to us in liquids and gases, whereby hot, less dense material rises and cooler, denser material sinks. It is the process that drives the great conveyor belt of ocean circulation and the rapid turmoil of winds in the

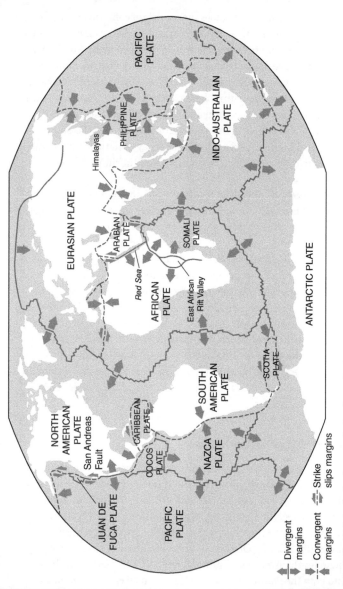

6. **Principal tectonic plates and plate boundaries.**

atmosphere, as much as it does the cooling of coffee in our mug or the rising of smoke up a chimney. At conditions of extremely high temperatures and pressures, however, the apparently 'solid' rocks of the outer Earth can behave as an extremely viscous 'fluid' that creeps or flows and thereby allows convection to occur. Even after 4.5 billion years, the natural heat contained within the core of the Earth, following its accretion and solidification, is still the principal heat engine driving convection in the mantle. Some heat, in addition, is derived from the decay of naturally occurring radioactive elements such as uranium and thorium.

Several methods now exist for calculating the rate of plate motion. Most reliable for present-day plate movement are direct observations made using satellites and laser technology. These show that the Atlantic Ocean is growing wider at a rate of between 2 and 4 centimetres per year (about the rate at which fingernails grow), the Indian Ocean is attempting to grow at a similar rate but is being severely hampered by surrounding plate collisions, while the fastest spreading centre is the East Pacific Rise along which ocean crust is being created at rates of around 17 centimetres per year (the rate at which hair grows). For the Pacific Ocean, however, the plates involved are being even more rapidly consumed at convergent plate boundaries, in a process known as subduction. The net result is, in fact, a reduction in size of the Pacific relative to the other oceans.

Backbone of the oceans

One of the most remarkable features of the oceans is an almost continuous range of submarine mountains that stretches for tens of thousands of kilometres across the face of the Earth. From the Arctic Ocean, almost directly beneath the North Pole, this mountain range runs axially down the length of both the North and South Atlantic Oceans. It links with another that completely encircles the Antarctic continent, from which great spurs divide into both the Pacific and Indian Oceans. In all cases, the mountains rise up to 3 kilometres above the adjacent ocean basins to within 2.5 kilometres of the ocean surface and, at their broadest, boast a width of over 1000 kilometres. This is the global mid-ocean ridge system—the backbone of the oceans.

Mid-ocean ridges are the sites at which new crust is continually being created as outpourings of a dark volcanic rock called basalt. The entire ridge system is essentially a linear belt of active volcanoes along which the majority of the Earth's internal heat is lost. Shallow-focus earthquakes occur in abundance as volcanic material is forced up to the surface, new crust is accreted, and the seafloor of yesterday pushed aside.

Hot magma wells up beneath the axis of the ridge driven by gigantic convection cells in the Earth's interior. A hot mush of crystals and molten rock is held in a magma chamber only a few kilometres beneath the seafloor, from which sheets and pipes (dikes) of magma force their way through the weakened crust above and erupt repeatedly on to the deep-ocean floor. Eruption beneath 2.5 kilometres of water is not violent but more like squeezing toothpaste out of a tube. The creation of new seafloor in this way has been captured on film, the dark, oozing basalt forming a series of small rounded domes known as pillow lava. The rind of these pillows cools very rapidly into a black glass as lava at over 1000°C meets near-freezing water.

Measuring the age of ocean crust yields an intriguing pattern on the seafloor, and recognition of the systematic increase in age away from the ridge crest provided the final proof of plate tectonic theory in the early 1960s. It is a quite remarkable story of how the Earth's magnetic field interacts with cooling magma, and the seafloor acts as a magnetic tape recorder. Iron-rich minerals crystallizing out from the hot lava become individually magnetized and oriented towards magnetic north, setting firmly as the lava solidifies. The present direction of the magnetic field is referred to as 'normal' and the opposite direction as 'reverse'. During the geologic past, the Earth's magnetic field has switched between normal and reverse with a variable periodicity, from many thousands to millions of years. Towing sensitive magnetometers behind research ships allows us to measure the sense of magnetization of crustal rocks. This reveals a pattern of elongate, narrow bands of normal and reverse magnetic direction with almost perfect symmetry on either side of the ridge crest. The ages of each band can be worked out using radiometric dating of the lavas and a complete map of the age of ocean crust constructed. This shows that the oldest crust was formed around 180 million years ago and lies furthest from the present-day spreading centre—the ridge crest. It also allows us to calculate the rate of spreading.

Black smokers, ophiolites, and hot spots

Along the mid-ocean ridge system, hydrothermal vents known as black smokers spout billowing plumes of hot, mineral-laden water. They are the result of cold seawater that percolates into cracks and fissures associated with seafloor spreading, then penetrates several kilometres into newly formed ocean crust and leaches out a cocktail of chemicals as it seeps downwards. When the water encounters hot magma it is forced back up to the surface, emerging on the seafloor as hot springs. Dissolved minerals rapidly precipitate as these often superheated waters (300°–400°C) come into contact with near-freezing temperatures at the ocean bottom. Their precipitates produce massive ore bodies rich in iron, copper, zinc, and a host of other metals, appearing almost like a cityscape of chimneys, spires, and stacks rising from the seafloor, or as pools of metal-rich sediments ponded in hollows along the mid-ocean ridge. Despite the high temperatures, pitch blackness, and a lethal concoction of toxic metals around the vents, a weird and wonderful flora and fauna thrive in this hostile environment.

Under the powerful forces involved in plate collision, fragments of ocean lithosphere can become detached and emplaced on continental crust. These fragments are known as ophiolites, and recognized by their distinctive suite of oceanic rocks. They are found in many orogenic (mountain) belts that stretch across the continents. Pillow lavas that were once the floor of a 2-billion-year-old ocean form part of a much fragmented ophiolite exposed in northern Quebec. More complete ophiolites are known from the island of Cyprus—an 80-million-year-old fragment of the now vanished Tethys Ocean—and, younger still, from many parts of the Japanese Island Arc. In such places, we can actually walk on the floor of a vanished ocean or stand astride the famous Mohorovičić or 'Moho' discontinuity between crust and mantle rocks. These are unforgettable experiences for an ocean scientist.

Another shrine for scientists who study the deep seafloor is Iceland. This volcanic island is part of the Mid-Atlantic Ridge, built up over millions of years from vast outpourings of oceanic lava, so that the crest of the ridge itself is exposed on land. Walking along the Thingvellir graben near Reykjavik is like a stroll down the narrow valley that runs along the axis of all slow-spreading mid-ocean ridges. Iceland has been formed directly above a hot spot, where molten material rises from the mantle in a narrow plume that extends as much as 600 kilometres beneath the surface. This hot spot was initiated around 63 million years ago beneath the continental crust that formerly joined Greenland with northern Europe. It was instrumental in the ensuing continental break-up and formation of a new spreading centre. Unusually for hot spots, it has remained active in the same place ever since.

Collision, subduction, and slippage

New crust is continually being added to the oceans all along the mid-ocean ridges, which implies that an equal volume of material must be destroyed annually. Destruction of ocean crust takes place in the world's largest recycling plants—the subduction zones, located along convergent plate margins. Because the rock material that makes up oceanic plates is denser than that of the continents, where the two collide the ocean plates are pushed under, or subducted. The parts of the seafloor where this occurs are commonly marked by deep trenches into which the oceanic lithosphere is thrust downwards to be remelted and recycled. Subduction results in earthquakes, volcanoes, island formation, and mountain building.

In the upper parts of a subduction zone, melting at the surface of the descending plate occurs as the result of tremendous frictional resistance. The wet sediments and rocks carried down on the plate surface as well as those of the deep lithosphere begin to melt at depths of 50 to 100 kilometres, where temperatures have risen to between 1200 and 1500 degrees Celsius. Localized pockets or chambers of molten rock (magma) are created. Some of this magma rises as intrusions into the lower crust where it cools and crystallizes as granite plutons—the principal rock type of the continents. Some of the molten material forces its way right to the surface in highly explosive volcanic eruptions.

The entire Pacific Ocean is rimmed by subduction zones and these are the location of the largest number of earthquakes and volcanoes anywhere in the world—hence its nickname the *Ring of Fire*. The tectonic plates that make up this ocean are constantly being recycled beneath the continents. The Nazca plate has been plunging beneath South America for at least 200 million years—the imposing Andes, the longest mountain chain on Earth, is the result. Further north, the small Juan de Fuca and Cocos plates are the only remnants of the once mighty Farallon plate, which floored much of the ancestral Pacific Ocean.

Along large stretches of the western and northern margins of the Pacific, subduction has led to the birth of a string of island volcanoes. Groups of volcanic islands occur some hundred kilometres behind the linked trench system, and separated from the continent by a broad marginal sea or completely isolated in the broad expanse of ocean. These groups in fact follow great curvilinear traces—known as island arcs—of which the Aleutian, Japanese, Philippine, and Indonesian arcs are some of the best known. With time, the once distinct islands amalgamate and landmasses grow to resemble small fragments of continent.

Mountain building that has occurred as the result of the subduction of oceanic crust beneath continental crust is well illustrated by the Andes Mountains. These have been forming for at least the past 100 million years from melting of lithospheric material subjected to intense frictional heating in the great subduction factory that underlies this imposing mountain range. As rocks melt at great depth so the molten magma begins to push upwards into the cooler rocks above, where the magma cools once more and begins to crystallize. Certain minerals, such as quartz and some feldspars, are the last to crystallize and so continue to rise with the remaining magma until they also crystallize out to form granite—the most common rock of the continents. All mountain chains are intruded by *plutons* of granite in this way, and where the magma continues right to the surface it will erupt as *andesitic* and *rhyolitic* lavas. The deep granite roots of the mountain chain are still buoyant, so that parts of the Andes are rising at rates of over 5 centimetres a year.

The mountain chains of Europe and the Middle East—the Alps, Apennines, Pyrenees, Betics, Hellenides, and Carpathians—have all formed in this way as Africa continues to plough northwards into Eurasia. The twists and turns of microplates, fragmented at the leading edge of collision, as well as island arcs, volcanoes, and abandoned trenches, make the jigsaw puzzle of mountain ranges and remnant basins in the whole Mediterranean region one of the most complex on Earth. In this sort of complexity, parts of ocean crust become isolated and scraped off on to the continent as ophiolites, while whole fragments of continent are detached and thrust high up over the mountains as *nappes*.

The chaotic and deformed rocks associated with such features are known as *melanges*.

In some regions, plates grind and slide past one another, rather than colliding or separating. These are known as transform boundaries in the oceans or strike-slip boundaries on land. Although this process conserves plate material it is by no means passive. The stepped offsets along mid-ocean ridges are examples of transform faults, some of which extend for many hundreds of kilometres either side of the ridge crest. The San Andreas Fault in California and the Southern Alpine Fault in New Zealand are examples of strike-slip boundaries on land. Along these giant fault lines, the two parts of a continent crunch, slip, and grate past one another, resulting in much earthquake activity.

Oceans past and present

We are so familiar with the map of our world, the shapes of continents and the breadth of oceans, that it is easy to believe in its permanence. In fact, the opposite is more true—continents have moved and grown, oceans have come and gone, ocean crust has been lost and remelted. These changes, though slow even by geological standards, have often had the most dramatic effects on Earth's climate, ocean circulation, floral and faunal distribution, and mass extinctions. It is therefore an important challenge in our science to be able to reconstruct the disposition of past seas and land areas accurately. The study of past oceans, known as palaeoceanography, is currently a 'hot topic' for research throughout the world.

There have been many cycles of change and many different oceans in the 4.5 billion years since planet Earth was formed. But our knowledge of those more distant in time is very sketchy. We can be much more confident in charting the past 250 million years of this story. This is roughly the time it has taken to create and then destroy one great ocean—the Tethys—and thence to shape the oceans of today (Figure 7).

At the start of this period, there was a single supercontinent known as *Pangaea* and a super-ocean known as *Panthalassa*. The large body of water that straddled the palaeo-equator as an arm of Panthalassa was known as the *Tethys Ocean*. Both poles were under water, global climate was generally warm, sea level was very low, and the continental interior extremely arid. Vast outpourings of volcanic lava, some of which now form the Siberian Traps, altered the chemistry of atmosphere and oceans. This combination

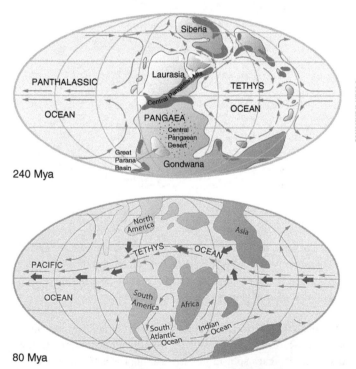

7. **Past oceans of the world: 250 million years ago to present.**

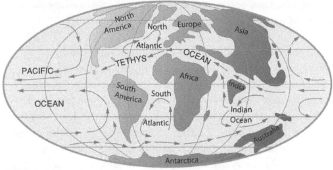

45 Mya

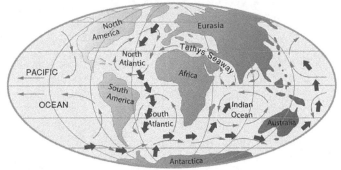

15 Mya

7. Continued.

of factors led to the greatest mass extinction of all time, at the end of the Permian Period, when an estimated 95 per cent of all living species (almost all marine) were wiped out. It was the Tethys Ocean that gradually nurtured life back to the world.

Pangaea only remained intact for a few tens of millions of years at most before it began slowly to break up and drift apart. In the first place Pangaea split more or less along the equator so that the Tethys Ocean then separated two large continents, Laurasia to the north and Gondwana to the south. The climate began to change, as moisture from this enlarged ocean penetrated the continental interior. Where there had been only desert, plant life spread and flourished. Along the margins of the continents, shallow shelf seas became home to coral reefs and tropical lagoons, sporting myriad new forms of marine life—coiled ammonites, squid-like belemnites, starfish, giant oysters, sea lilies, and fishes, preyed upon by ichthyosaurs, pliosaurs, and the ancestors of modern sharks.

By around 120 million years ago, South America and Africa began to drift apart and the South Atlantic was born. It is this era that is known as the true *Age of the Ocean*, for sea levels rose higher than at any time during the past billion years, perhaps as much as 350 metres higher than today. Only 18 per cent of the globe was dry land—82 per cent was under water. These excessively high sea levels were the result of increased spreading activity—new oceans, new ridges, and faster spreading rates all meant that the mid-ocean ridge systems collectively displaced a greater volume of water than ever before. Global warming was far more extreme than today. Temperatures in the ocean rose to around 30°C at the equator and as much as 14°C at the poles. Ocean circulation was very sluggish.

In these extreme conditions the oceans became dangerously low in oxygen. Several periods of near oxygen starvation occurred, globally or at least semi-globally. These are known as *ocean anoxic events*, during which large quantities of undecayed organic matter were preserved in seafloor sediments. Wherever we have drilled into sediments of this age beneath the ocean floor or, indeed, where rocks of this age are now exposed on land, we encounter dark grey rocks called black shales. Over half the world's oil reserves have been generated from these rich source rocks. There then followed a long period of warm, tranquil seas dominated by a profusion of plankton and a diverse animal life. This stability ended dramatically 65 million years ago in another mass extinction event, in which some

50 per cent of known species disappeared—including the dinosaurs on land.

Sea levels fell, global cooling set in, and the continents continued to break up. Antarctica separated from Australia, and drifted across the South Pole, taking up more or less its present position. India had already detached and begun a slow northward drift. Between 60 and 50 million years ago the leading edge of this detached Indian plate began to collide with the northern landmass of Laurasia. Before long, a full continent–continent collision ensued and a new mountain range began to emerge—the Himalayas. This is Earth's highest mountain chain, claiming all ten of the world's peaks in excess of 8000 metres. It even has parts of the early Tethys Ocean floor, thrust up and over the rising mountains, and exposed at the surface today—for example, at the 8500 metre summit of Mount Makalu, a neighbour of Everest on the border between Nepal and Tibet.

As India collided with China, so Africa closed in on central and western Eurasia and South America drifted towards its northern counterpart. The Tethys Ocean slowly disappeared, and only slithers of rock from its former floors are now caught up in the Himalayas, in China, and in the Middle East. In a final twist of its fate, the last remnants of the Tethys became an irregularly shaped landlocked water body roughly similar to the present-day Mediterranean and Black Sea at about 6.5 million years ago. This remnant ocean repeatedly evaporated to dryness and then reflooded over a period of one million years, leaving behind an extensive layer of evaporite sediment (mainly gypsum and halite) that is up to 2000 metres thick in places. Eventually a stable connection with the Atlantic Ocean was established through the Straits of Gibraltar and normal marine conditions ensued.

Chapter 3
Ocean floor

The jigsaw puzzle of tectonic plates and their movement about the Earth determines the nature and shape of the seafloor. This is as varied as it is variable—boasting the world's longest mountain chain and its deepest chasm, valleys that would swallow the Grand Canyon, and vast lifeless plains twice the size of Sahara. But all these features change with time and patterns of change are forever cyclic, though the time scales vary and the past is never exactly repeated. The plate tectonic cycle, as outlined in Chapter 2, is the longest of these, and closely linked with the cycle of the rocks, from continent to ocean to the Earth's interior and back. The flux of sedimentary materials into and through the oceans is complex, at times sudden and catastrophic, otherwise slow and continuous. This chapter examines the morphology of the ocean floor, its composition, sediment cover, and cycles of change.

In the late 1980s, I led an international scientific expedition with Jim Cochran (of Lamont-Doherty Geological Observatory in the USA) for which we took the *JOIDES Resolution* scientific drillship to the middle of the Indian Ocean. Having surveyed the nature of the seafloor and what lay beneath, using seismic imaging, we selected three representative sites and drilled into the most distant part of the world's largest deep-sea fan (the Bengal Fan). This gigantic sedimentary deposit, which extends 3000 kilometres from the Ganges Delta to somewhere south of the

equator at a water depth of over 5000 metres, has been growing
ever since the Himalayas began to push upwards. Our deepest
hole penetrated 1000 metres below the seafloor, reaching back to
sediment eroded from the Himalayas some 18 million years ago.
By analysing the composition of the sediment column above
this, we unravelled a picture of progressive denudation of this
mountain range. Today, what was originally the deep core of the
Himalayas is exposed at its summit. We also found that the
sediment cover and the ocean crust below were being slowly
buckled and deformed into a regular pattern of enormous folds,
as a result of the intense pressure built up as the Indian plate
continues to push into the Eurasian plate. The ocean floor is
varied and complex.

Shape of the seafloor

The land–ocean boundary is known as the shoreline. Seaward
of this, all continents are surrounded by a broad, flat continental
shelf, typically 10–100 kilometres wide, which slopes very gently
(less than one-tenth of a degree) to the shelf edge at a water
depth of around 100 metres. Beyond this the continental slope
plunges to the deep-ocean floor. The slope is from tens to a few
hundred kilometres wide and with a mostly gentle gradient
of 3–8 degrees, but locally steeper where it is affected by faulting.
The base of slope abuts the abyssal plain—flat, almost featureless
expanses between 4 and 6 kilometres deep. The oceans are
compartmentalized into abyssal basins separated by submarine
mountain ranges and plateaus, which are the result of submarine
volcanic outpourings.

Those parts of the Earth that are formed of ocean crust are
relatively lower, because they are made up of denser rocks—basalts.
Those formed of less dense rocks (granites) of the continental
crust are relatively higher. Seawater fills in the deeper parts, the
ocean basins, to an average depth of around 4 kilometres. In fact,
some parts are shallower because the ocean crust is new and still

warm—these are the mid-ocean ridges at around 2.5 kilometres—whereas older, cooler crust drags the seafloor down to a depth of over 6 kilometres. The mid-ocean ridges are cut through by deep, almost linear gashes known as fracture zones, which can be over 100 kilometres long. Where ocean crust plunges into long, narrow trenches at subduction zones, the floor of the ocean can drop away to a water depth of over 9 kilometres—and a maximum of 10,924 metres in the Challenger Deep of the western Pacific Marianas Trench.

Isolated volcanic seamounts rise from the deep seafloor, in some cases piercing through the sea surface to form small remote islands, such as the many favourite holiday destinations scattered across the Pacific and Indian Oceans.

Seafloor sediments

The seafloor is almost entirely covered with sediment. In places, such as on the flanks of mid-ocean ridges, it is no more than a thin veneer. Elsewhere, along stable continental margins or beneath major deltas where deposition has persisted for millions of years, the accumulated thickness can exceed 15 kilometres. These areas are known as sedimentary basins—the repository of much hidden wealth, both economic and historical.

All sediments are made up of discrete particles of material that accumulate with time in distinct layers at the surface of the Earth (Table 2). Some are deposited on land, most at sea. With the passage of time and continued accumulation, older sediment is buried to greater and greater depths. Under the high temperatures and pressures that result from burial, individual grains are compacted ever more tightly until, at a burial depth of one or more kilometres, they begin to fuse together. Pore fluids are squeezed outwards and upwards to escape the ever-increasing pressure, and certain minerals precipitate out and act as a cementing agent binding together the once separate particles.

Table 2 Principal rock types

Igneous Rocks

Origin	Cooling and crystallization of molten rock (magma)
Examples	Plutonic rocks: granite, diorite, gabbro
	Volcanic rocks: rhyolite, andesite, basalt, pumice, obsidian

Metamorphic Rocks

Origin	Recrystallization of pre-existing rocks at high pressures and temperatures
Examples	Slate, schist, quartzite, hornfels, gneiss, marble

Sedimentary Rocks

Origin	Deposition of particulate material from pre-existing rocks, organisms, and dissolved chemical salts
Terrigenous	Land-derived, from erosion and weathering of pre-existing rocks
	Examples: sandstone, conglomerate, mudstone, shale
Biogenic	Biological origin, from broken up fragments of organisms
	Examples: limestone, chert, phosphorite, coal
Chemogenic	Chemical origin, from precipitation of dissolved salts in seawater
	Examples: halite, gypsum, manganese nodules, ironstone, tufa, travertine
Volcanogenic	Volcanic origin, from fragmented volcanic/igneous rocks
	Examples: volcanic ash, volcanic tuff, pyroclastite, ignimbrite

Many different types of sedimentary material find their way into the ocean basins. Terrigenous material is derived primarily from the physical and chemical weathering of pre-existing rocks—igneous, metamorphic, and sedimentary alike. This debris is delivered to the ocean via rivers, coastal erosion, floating ice, and as wind-blown dust and yields gravel, sand, silt, and clay (mud) deposits common throughout the oceans. Upon lithification these sediments become conglomerates, sandstones, and mudstones.

Biogenic material, produced from living organisms, includes both hard shells and soft organic tissue. It is supplied to the seafloor from copious amounts of plankton in surface waters as well as the break-up of reefs and shell banks. Various types of limestone and a tough silica rock (chert or flint) are the end result. Where soft organic remains dominate, then black shale and coal are eventually formed.

Chemogenic particles are those derived from the wide range of dissolved chemicals in ocean water. A steady-state chemistry is maintained in the oceans, subject always to large material flux in and out of solution. Where these conditions change, then the more common chemicals are likely to precipitate out of solution. Coastal lagoons or shallow seas partially cut off from the global ocean can rapidly become salt pans if climatic conditions lead to an excess of evaporation over precipitation. Natural salts (halite) and gypsum are the result. Other physico-chemical conditions locally favour the precipitation of iron, manganese, and a host of other metals as nodules, crusts, and chimney edifices in the deep sea.

The map of sediment distribution in the global ocean reveals an intriguing pattern (Figure 8). Whereas some features are readily understood in terms of different sediment types and supply routes, other aspects are much less obvious. That the continental margins, including deltas and submarine fans seaboard of major rivers, are composed of terrigenous sediments is a clear reflection of this material being eroded and supplied from the continent. Likewise, glacigenic sediments clearly dominate high-latitude seas. But why is calcareous biogenic sediment more common in the central part of the Atlantic and Indian Oceans, and over certain broad tracts of the Pacific? And why are there three broad latitudinal bands of siliceous oozes? The answers lie in the complex nature of productivity and preservation in the oceans.

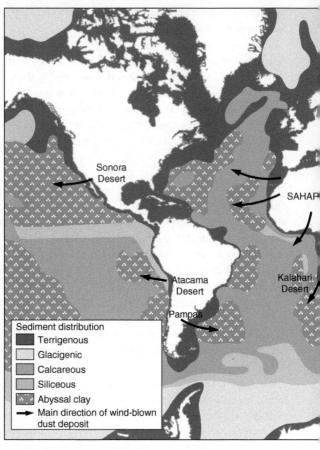

Sonora
Desert

SAHAR

Atacama
Desert

Kalahari
Desert

Pampas

Sediment distribution
■ Terrigenous
□ Glacigenic
▨ Calcareous
▥ Siliceous
▨ Abyssal clay
→ Main direction of wind-blown
dust deposit

8. Pattern of sediment distribution on seafloor.

Planktonic organisms (the result of primary productivity in the
surface waters) supply both calcareous and siliceous skeletal hard
parts as a steady rain to the seafloor. But at a certain depth in the
water column, currently around 4.5 kilometres in the Atlantic, the
bottom water becomes more acidic and dissolves any calcareous
material that sinks this far. Calcareous oozes, therefore, can only be
preserved where the depth of the seafloor lies above what is known

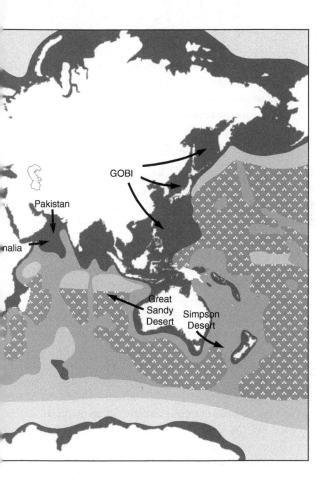

as the *carbonate compensation depth*, which explains their distribution over more elevated regions of the seafloor. Siliceous skeletons are even more soluble in seawater and will only reach the seafloor and be preserved as sediment if they are produced in sufficient quantities to completely swamp the system. This can occur beneath surface zones of especially high primary productivity—in a broad equatorial belt, and in two high-latitude belts around the world, as well as in shelf regions on the western margins of continents. In the deepest, most remote parts of the ocean basins, only abyssal red clay is deposited. This accumulates at rates of a few millimetres every thousand years, so that all traces of iron minerals have time to become fully oxidized on the seafloor yielding the characteristic rust colour of these abyssal clays.

Canyons, slopes, and deep-sea fans

The transfer of terrigenous sediment from the continents across the continental shelf and slope to the deep-ocean basins is part of the sedimentary cycle. Rocks are eroded under the chemical and physical influence of water and break into their constituent minerals and grains. These are washed by rivers towards lakes and the sea, rounding and grinding as they move. Ocean currents further polish and sort sediments across the continental shelves, while submarine slides and flows eventually transport sands and muds deep into the ocean basins. When sediment is deposited on the seafloor it still holds water, rather like a sponge, and it is the cycling of millions of litres of water through the pore spaces between individual particles that eventually leads to the precipitation of mineral cements and lithification into a hard sedimentary rock.

It is not surprising that the world's greatest mountain range, the Himalayas, feeds the muddiest rivers, and has built up two of the world's largest deltas—the Ganges and Indus Deltas (Figure 9). Beyond these, billions of tonnes of sediment are emptied into the deep Indian Ocean. The Bengal Fan, a gigantic submarine delta, is the world's single largest sediment body. It covers an area of over 1 million square kilometres and reaches a maximum thickness of at least 15 kilometres, thinning to a feather edge 3000 kilometres out into the Indian Ocean.

Four main types of slope make up the ocean margins, each characterized by different morphology and processes. The first is smooth and gentle, typically swept by the action of contour-hugging

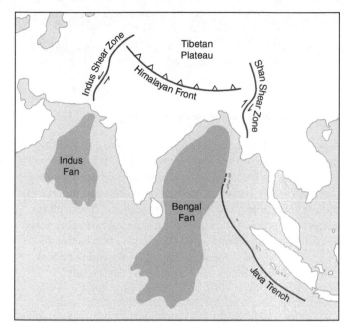

Ocean floor

9. Bengal and Indus submarine fans, north-east Indian Ocean.

bottom currents. The second type is steep and rugged, with
gradients over 10° and locally more than 45°. These are found
off coral reefs and carbonate banks, and also where major faulting
has accentuated the slope angle. The third type is irregular in
surface morphology and quite gentle (say 5–10° gradients). They
are characterized by submarine slides and mudflows as a result
of great instability due to earthquakes and volcanism. They occur
along active (convergent or transform) continental margins and
where there has been very rapid build-up of thick sediment piles
from glaciomarine or riverine supply. The fourth type is dissected
by channels of all different sizes, shapes, and styles.

These channels are the principal conduits for transporting
sediment to the deep oceans. In some cases we can see where they

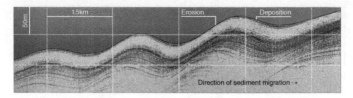

10. Seismic profile of deep-sea sediment waves.

are clearly linked to an onshore river system, and presume they were first carved when sea levels were much lower and the rivers flowed out across broad continental shelves. Off some of the major river deltas, even deeper troughs have been excavated through the pro-delta slope muds. The Swatch-of-No-Ground trough off the Ganges Delta is over 600 metres deep, while the main channel of the Laurentian Fan off eastern Canada is up to 800 metres deep and 15 kilometres wide. This could swallow the world's highest building, the Burj Khalifa, as well as its host city, Dubai!

Some of the channels are highly sinuous with tight meanders, whereas others are straight or gently sinuous. Where submersible dives have reached the bottom of such channels or core samples have been recovered from their narrow floors, coarse gravel and sand have often been observed. The channel levees or overbank regions, by contrast, are built up of thick, well-layered deposits of fine silts and clays, material that has spilled out of the channel during the passage of a giant flow—a submarine river flooding its banks. Some remarkable images across these channel overbank areas show huge fields of regularly spaced giant sediment waves (Figure 10). The wavelength is typically between 1 and 2 kilometres, while the amplitude is several tens of metres.

Hidden currents of the deep

Much is still to be learned about the processes that move material from the basin margins downslope, either through the many channel systems or across the slope by some other means. One important and common process on most slopes is that of submarine slides and slumps. It involves the sudden downslope displacement of the upper layers of sediment along a basal shear plane. Single slide masses range from very small localized displacements to very large bodies that may be as much as several hundred metres in thickness and over 100 cubic kilometres in volume. Examples of such catastrophic slides at sea have involved the complete loss of a drilling platform with all hands on board off the Mississippi Delta, the removal of half the new runway at Nice Airport in southern France and its sliding downslope into the Mediterranean Sea, and the overnight disappearance of a Ukrainian village from the Crimean Peninsula into the Black Sea. Such large-scale slides generally lead to the generation of major tsunamis and further disastrous effects around the periphery of semi-enclosed basins. The destruction of the Minoan civilization on Crete some 3500 years ago is believed to have resulted from a combination of explosive volcanism on the island of Santorini, followed by submarine sliding around the flanks of the caldera and the generation of devastating tsunamis.

During particularly chaotic slide events, much seawater is ingested by the moving mass so that it becomes much less coherent and more lubricated. Debris flows and turbidity currents are the result. Turbidity currents are believed to be one of the most common processes of the deep-sea environment although, paradoxically, a full-sized prototype has never yet been observed in nature. The process remains a phantom of the deep about which we have pieced together a great deal of information from their dramatic effects—the regular breaking of submarine telegraph cables, and their deposits—turbidites. The larger turbidity currents, generated from gigantic basin-margin slides, may reach over 500 metres in thickness, 10 kilometres in length, and speeds of over 70 kilometres per hour, and eventually deposit a 2-metre-thick layer of sand and mud over an area the size of France or the State of Texas, for example. Jim Cochran and I found exactly this type of mega-turbidite deposit when we drilled on the Bengal Fan in the central Indian Ocean.

Chapter 4
Chemical broth

The Earth is a single unity linked everywhere by oceans and seas. The remarkable properties of the simple water molecule have not only coloured the planet blue when viewed from outer space, but also allowed Earth to retain its hydrosphere and all its cycles, the seas to develop their saltiness and hierarchy of tiers, and life to develop and flourish. Water is a super-solvent, absorbing gases from the atmosphere and extracting salts from the land. About 3 billion tonnes of dissolved chemicals are delivered by rivers to the oceans each year, yet their concentration in seawater has remained much the same for at least several hundreds of millions of years. Some elements remain in seawater for 100 million years, others for only a few hundred, but all are eventually cycled through the rocks. The oceans act as a chemical filter and buffer for planet Earth, control the distribution of temperature, and moderate climate. Inestimable numbers of calories of heat energy are transferred every second from the equator to the poles in ocean currents. But, the ocean configuration also insulates Antarctica and allows the build-up of over 4000 metres of ice and snow above the South Pole. This chapter considers ocean chemistry and physics.

I am not a chemical oceanographer, but I do use the chemical and physical signature of seawater, captured in the composition of microscopic plankton, to decode and explain past changes in ocean temperature and circulation. As a PhD student of Dalhousie University in Canada, I spent many hours with a microscope counting the tiny coiled shells of *Globigerina pachyderma*—a particular species of planktonic foraminifera. Those that coiled to the right (dextral) lived in warm waters, as recovered from the surface sediments today. Those that coiled to the left (sinistral), recovered from deeper within my sediment cores, had lived in much colder waters of the past ice ages. I could therefore date the other changes seen in the sediments in relation to past glacial and interglacial climates.

Science has moved on (thankfully), so there is now a barrage of sophisticated chemical and other analyses that can be made on the skeletons of foraminifera, rather than counting their coiling direction. A few years ago, I sailed with Dave Hoddell of Cambridge University to drill a series of deep wells into the continental slope muds off south-west Portugal. Now, after tens of thousands of analyses made on samples of the sediments recovered in different laboratories across the world, we have the world's most complete marine record of climate change over the past 1.8 million years. This has only become possible because we have learned so much more about ocean chemistry, how it responds to climate change, and how it influences the chemical make-up of marine life.

Why the sea is salty

It is perhaps unsurprising that the substance that makes the Earth unique has a very simple and well-known chemistry. Water (H_2O) is made up of two parts hydrogen, the most abundant element in the universe, and one part oxygen, an element that accounts for nearly half the weight of the Earth's crust. But the way in which these elements combine yields a common substance with rare and remarkable properties that are crucial in determining physical and chemical conditions across the surface of the world.

The oceans formed early in the planet's history as water escaped from the crust and mantle through the process of outgassing and as a rain of cometary particles bombarded Earth. The composition of the first seas was mostly one of freshwater together with some dissolved gases. Today, however, the world ocean contains over 5 trillion tonnes of dissolved salts, and nearly 100 different chemical elements, including some 5.5 billion kilograms of gold. If the oceans' water evaporated completely, the dried residue of salts would be equivalent to a 45-metre-thick layer over the entire planet. Marine chemists aim to understand the nature of ocean salinity and the flux of elements into and out of solution. The chemical compounds that form the building blocks of life—carbon, nitrogen, phosphorus, hydrogen, and oxygen, as well as several essential trace elements—all occur in water. How and when did they get there? And do they remain in constant proportions? The global cycling of these elements is both extremely complex and vital for life to continue to flourish.

The super-efficiency of water as a solvent is due to an asymmetrical bonding between hydrogen and oxygen atoms (Figure 11). The resultant water molecule has an angular or kinked shape with weakly charged positive and negative ends, rather like magnetic poles. This polar structure is especially significant when water comes into contact with substances whose elements are held together by the attraction of opposite electrical charges. Such ionic bonding is typical of many salts, such as sodium chloride (common salt) in which a positive sodium ion is attracted to a negative chloride ion. Water molecules infiltrate the solid compound, the positive hydrogen end being attracted to the chloride and the negative oxygen end to the sodium, surrounding and then isolating the individual ions, thereby disaggregating the solid. An apparently simple process, but extremely effective.

The rain that falls on land as part of the water cycle, from the very first that ever fell on Earth to the present time, actively dissolves a multitude of chemicals from the rocks. Those elements that are

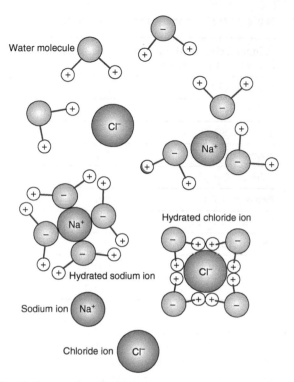

Water molecule

Cl⁻

Na⁺

Hydrated chloride ion

Na⁺

Hydrated sodium ion

Sodium ion Na⁺

Chloride ion Cl⁻

11. Chemical properties of water.

easily disaggregated, rather than the more abundant, are leached in greater amounts. Certain gases in the atmosphere, notably carbon dioxide and sulphur dioxide, are dissolved in the water droplets as they fall, so forming weakly acidic rain. This is an even more effective solvent than pure water. Over many aeons, the oceans slowly accumulated dissolved chemical ions (and complex ions) of almost every element present in the crust and atmosphere. Outgassing from the mantle from volcanoes and vents along the mid-ocean ridges contributed a variety of other elements, including juvenile (new) water. Most elements, apart from hydrogen and oxygen, are present in extremely small

Table 3 Major constituents of seawater

Principal chemical solute and concentration in parts per thousand

Chloride (Cl$^-$)	19.3
Sodium (Na$^+$)	10.7
Sulphate (SO$_4^{2-}$)	2.7
Magnesium (Mg^{2+})	1.3
Calcium (Ca^{2+})	0.41
Potassium (K$^+$)	0.38
Bicarbonate (HCO$_3^-$)	0.14
Bromide (Br$^-$)	0.065
Strontium (Sr^{2+})	0.013
Fluoride (F$^-$)	0.001

amounts (parts per thousand, ‰)—these include sodium, chlorine, magnesium, sulphur, calcium, potassium, and carbon. The next five most common are measured in parts per million (ppm), whereas all the rest are present only as trace elements.

The total concentration of dissolved (inorganic) substances in water is referred to as its salinity. This varies from about 3.3 per cent to 3.7 per cent by weight (or 33–37‰), depending on local factors such as precipitation, evaporation, and riverine influx. The average is 3.5 per cent (35‰) and has remained more or less constant for as long as measured records are available and, in fact, has changed little for the past 200 to 300 million years (Table 3). But this poses a conundrum—why doesn't the ocean get progressively saltier as a result of continued riverine input and outgassing from mid-ocean ridges? There is also a related puzzle—why is the chemistry of river water so different from that of seawater? Rivers typically contain a much larger proportion of rock-derived elements (silicon, calcium, magnesium, iron, and aluminium).

Cycles and sinks

The abundance of water on Earth and its occurrence in multiple forms is unique in the solar system. The storage of water and its transfer between different forms and reservoirs is known as the hydrological cycle. This is the single largest sink for incoming solar radiation, and has the greatest effect on life, climate, and the shape of the land. The dominant reservoir is the ocean and its marginal seas, housing over 90 per cent of Earth's water. The second principal reservoir is subsurface water—some held in soils and rocks, but most residing in the tiny pore spaces of sediment that still remain open down to several kilometres beneath the seafloor. Ice sheets, glaciers, and snow hold most of the world's freshwater, with only a very small fraction cycling through the atmosphere, rivers, and lakes.

Water everywhere is constantly on the move. Solar energy causes large-scale evaporation from the oceans, lakes, and rivers, as well as evapotranspiration from plants on land. Atmospheric winds disperse water vapour across the globe until it cools and condenses, falling back to the surface as rain or snow. Some 80 per cent of the rain falls directly back into the ocean, the rest eventually finding its way back via rivers and groundwater flow. The average time a single molecule of water remains in any one reservoir varies enormously. It may survive only one night as dew, up to a week in the atmosphere or as part of an organism, two weeks in rivers, and up to a year or more in soils and wetlands. Residence times in the oceans are generally over 4000 years, and water may remain in ice caps for tens of thousands of years.

Although the ocean appears to be in a steady state, in which both the relative proportion and amounts of dissolved elements per unit volume are nearly constant, this is achieved by a process of chemical cycles and sinks. The input of elements from mantle outgassing and continental runoff must be exactly balanced by their removal from the oceans into temporary or permanent sinks. The principal sink is the sediment and the principal agent removing ions from solution is biological.

Many elements are removed directly into the sediment by absorption on to clay particles as they fall through the water column or lie on the seafloor, and by the chemical precipitation of ferro-manganese nodules, black-smoker vents, and evaporite sediments along arid shorelines. As sediment lithifies into rock, more chemical elements are bound up as cement. Some sediment as well as water is lost back into the mantle at subduction zones, much of which is subsequently uplifted into mountain ranges on land, so that the cycle of weathering, dissolution, and runoff begins again.

There are now sufficient data that we can build up a semi-quantitative measure of the rates at which different elements are added to or removed from the ocean reservoir. It is therefore possible to calculate the average length of time an element spends in the ocean (its residence time) from the relationship:

$$\text{Residence time} = \frac{\text{Amount of element in the ocean}}{\text{Rate of addition / removal from the ocean}}$$

The residence times of different elements vary enormously from tens of millions of years for chloride and sodium, to a few hundred years only for manganese, aluminium, and iron. Ocean water itself has a residence time of around 3500 years, as determined from measured rates of evaporation, precipitation, and runoff. As the Earth's oceans are more than 4 billion years old, we can infer that, on average, individual water molecules have cycled through the atmosphere (or mantle) and returned to the seas more than a million times since the world ocean formed.

The data on chemical residence times are of great practical significance. As pollution of the global ocean accelerates, with ever-increasing volume and variety of anthropogenic input—lead from petrol, mercury, and other heavy metals, radioactive elements, pesticides, petroleum, plastics, sewage, and acid rain—it is important that we know the effects this will have and the timescales involved. Natural oil seepage was once the dominant input of immiscible petroleum to seawater, mostly finding its way quickly back into the sediment or washed up along the shoreline. Transport and waste discharge now equal or exceed the natural supply, but the ultimate sink remains the same. Deep ocean floor sediments are actively being considered as safe ultimate repositories for radioactive wastes and other highly toxic products. But research shows clearly the active cycling of water and chemicals through the sediments and the potential for rapid leaching from burial waste. All these aspects need further investigation.

One of the most fundamental exchanges between the ocean and atmosphere, from the viewpoint of life on Earth, is that of carbon dioxide. The plants and animals that live in the ocean are very sensitive to the chemistry of seawater, including the amount of dissolved carbon dioxide. This is more than sixty times that in the atmosphere. All marine plants (as well as land plants) take in carbon dioxide during the process of photosynthesis and release oxygen. During animal respiration, the reverse is true, so that all animals take in oxygen and breathe out carbon dioxide. From natural sources, vast quantities are delivered to the oceans daily as rainfall. Human pollution, principally from burning fossil (carbon-based) fuels, provides additional input to the oceans of an estimated 3 billion tonnes of carbon dioxide per year.

Any large excess of acid rain falling into freshwater rivers and lakes rapidly increases their acidity until they become barren of life. However, the effect of that falling at sea is moderated due to a series of reactions involving carbon dioxide and the water molecule that serve to buffer the effects of any major increase or decrease in the carbon dioxide content of the oceans. Until recently, these processes have maintained seawater as a mildly alkaline solution, with an average pH value of 7.8 and a range generally between 7.5 and 8.4. There is now a serious trend towards ocean acidification (lower pH values) as a direct result of the increased atmospheric concentration of carbon dioxide. The weakly alkaline nature of seawater is very beneficial for plants and animals that make carbonate shells, as more acidic solutions readily dissolve carbonate, and also for the formation of many complex compounds that make up living organisms.

Ocean temperature

The remarkable physical properties of water are even more fundamental to life on Earth than its chemical efficiency as a super-solvent. They prevent extreme temperature ranges over much of the planet by acting as a global thermostat, inhibit instantaneous vaporization of the oceans, engage in all life processes, and ensure that water is the dominant component of living organisms.

Because of its polar structure and hydrogen bonding between individual molecules, water has both a high capacity for storing large amounts of heat and one of the highest specific heat values of all known substances. This means that water can absorb (or release) large amounts of heat energy while changing relatively little in temperature. Beach sand, by contrast, has a specific heat five times lower than water, which explains why, on sunny days, beaches soon become too hot to stand on with bare feet while the sea remains pleasantly cool.

Solar radiation is the dominant source of heat energy for the ocean and for the Earth as a whole. The differential in solar input with latitude is the main driver for atmospheric winds and ocean currents. Both winds and especially currents are the prime means of mitigating the polar–tropical heat imbalance, so that the polar oceans do not freeze solid, nor the equatorial oceans gently simmer. For example, the Gulf Stream transports some 550 trillion calories from the Caribbean Sea across the North Atlantic *each second*, and so moderates the climate of north-western Europe. Any small change in the global atmosphere–ocean regime induced by human or natural causes can have a significant impact on heat transfer and therefore on regional climate.

A dramatic natural example of this impact was the opening of the Drake Gateway as South America drifted away from Antarctica, coupled with the opening of the Tasman Gateway south of Australia. Both these events occurred around 25–30 million years ago, allowing the onset of the Antarctic Circumpolar Current and hence complete isolation of the Antarctic continent from the warming effects of the tropical ocean. This was probably the principal cause of prolonged global cooling that led to an icehouse Earth and the recent ice ages.

Colours of the ocean

From space, both the atmosphere and ocean are mainly blue, apart from reflective white clouds and brown-yellow plumes of river-borne sediment near the coast. But for anyone who has spent time at sea or indeed walking along its extensive coastline, the sea in reality is a chameleon of colour change. But first, why is it mostly blue?

The sunlight incident on the sea has a full spectrum of wavelengths, including the rainbow of colours that make up the visible spectrum (Figure 12). The longer wavelengths (red) and very short (ultraviolet) are preferentially absorbed by water, rapidly leaving near-monochromatic blue light to penetrate furthest before it too is absorbed. The dominant hue that is back-scattered, therefore, is blue. In coastal waters, suspended sediment and dissolved organic debris absorb additional short

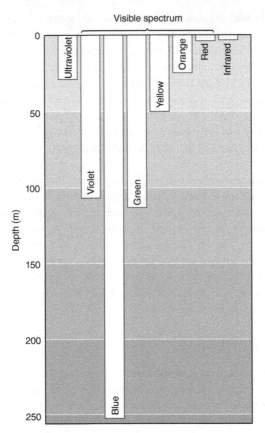

Visible spectrum

12. Underwater light.

wavelengths (blue) resulting in a greener hue. The crystal lattice
structure of ice behaves in a similar way, yielding the green-coloured
cliff faces of Antarctic ice shelves and floating icebergs.

Microscopic phytoplankton in surface waters utilize sunlight
for photosynthesis. But, as sunlight of all wavelengths is rapidly
absorbed, even in the clearest waters, photosynthesizing organisms

cannot live below about 100 metres. This is the photic zone, which may be as little as 20 metres in near-shore waters. There is also a twilight zone in the open ocean, with the barest hint of energy from the Sun penetrating up to 750 metres below the surface. Below that the ocean is pitch black, except for scattered bioluminescence generated by the organisms that inhabit this underworld.

Underwater sound

Marine mammals communicate with sound—well known as the often quite complex language and song of cetaceans. A range of invertebrates also make sounds, some crustaceans (such as the mantis shrimp) are known to click their limbs, and cod can grunt. Most fish have lines of sensors along their bodies that pick up vibrations from sound and movement, and many also create sound themselves.

The speed of sound in seawater is about 1500 metres per second, almost five times that in air. It is even faster where the water is denser, warmer, or more salty and shows a slow but steady increase with depth (related to increasing water pressure). There is a zone of minimum sound velocity at around 1000 metres depth, known as the *deep-sound channel* (or SOFAR). Sound generated in this channel is focused by refraction from above and from below and so loses little energy due to dispersion. Whales use this channel to communicate over great distances, and scientific field tests have shown that sound waves can travel for 25,000 kilometres from Australia to Bermuda.

We now have a wide array of scientific instruments for generating and directing sound impulses of different frequencies, and receivers for recording and processing the signals that return. These underwater sonar devices first produced a simple record of bathymetry, usually as a single line beneath the ship's track. Today we use sophisticated, sideways-looking, three-dimensional imagery to map great swathes of seafloor, pushing resolution down to the scale of sediment waves and dunes, submarine telegraph cables, and pipelines. It is these techniques that have been used to survey 120,000 square kilometres of seafloor in the remote south Indian Ocean as part of a coordinated international effort to locate remains of the Malaysian Airlines Flight 370 that went missing in 2014. It is the very first time that most of this area has ever been surveyed. Exploration geophysicists have become equally sophisticated in using sound to look through the water and into the sedimentary record beneath the seafloor in the search for oil and gas.

Even within the water column, we can use sound to provide important information. The Doppler principle—the change in pitch due to relative motion between the sound source and the receiver—is now used in acoustic Doppler current profilers. These yield measurements of current velocities accurate to within 1 centimetre per second for up to 128 different depth slices through a major ocean current up to 1 kilometre in thickness. Such studies have already greatly improved our understanding of otherwise 'invisible' internal waves, which travel along density interfaces within the ocean and break against the continental shelf-edge (Figure 13).

Ocean layers

From top to bottom, the ocean is organized into layers, in which the physical and chemical properties of the ocean—salinity, temperature, density, and light penetration—show strong vertical segregation. This layer-cake structure is always present but variable in its stability and subject to periodic breakdown and cyclic change (Figure 14).

Almost all properties of the ocean vary in some way with depth. Light penetration is attenuated by absorption and scattering, giving an upper photic and lower aphotic zone, with a more or less well-defined twilight region in between. Absorption of incoming solar energy also preferentially heats the surface waters, although with marked variations between latitudes and seasons. This

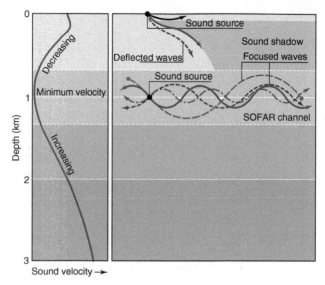

13. Underwater sound.

results in a warm surface layer, a transition layer (the thermocline) through which the temperature decreases rapidly with depth, and a cold deep homogeneous zone reaching to the ocean floor. Exactly the same broad three-fold layering is true for salinity, except that salinity increases with depth—through the halocline.

The density of seawater is controlled by its temperature, salinity, and pressure, such that colder, saltier, and deeper waters are all more dense. A rapid density change, known as the pycnocline, is therefore found at approximately the same depth as the thermocline and halocline. This varies from about 10 to 500 metres, and is often completely absent at the highest latitudes. Winds and waves thoroughly stir and mix the upper layers of the ocean, even destroying the layered structure during major storms, but barely touch the more stable, deep waters.

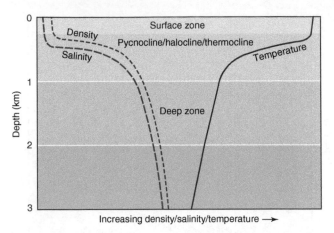

14. Ocean layers.

These deep waters constitute 80 per cent of the ocean and are made up of different water masses, most of which are generated at high latitudes by the cooling and sinking of polar waters, which then spread out at depth towards the equator. Depending on just where they are formed (Arctic, Antarctic, Labrador, or Greenland Seas, for example) and how much subsequent mixing has taken place, their properties differ. The coldest, most saline waters, formed around Antarctica, are the densest and hence form the deepest layer in the oceans. Above this, the layer-cake structure of the water column is complex and variable.

But the layer which is perhaps of greatest importance is also that which is most difficult to study, for it is a microlayer at the sea surface only a few millimetres thick. It is through this thinnest of horizons that 70 per cent of the Earth's solar energy is absorbed, most of the rainwater, carbon dioxide, and oxygen are exchanged, and enormous volumes of particulate matter and pollutants are passed. The consequent influence on the ocean–climate couple and on the whole of marine life below is highly significant.

Stagnant pools

One effect of layering is isolation of the deep sea from its well-mixed, light and airy surface layer. Plankton blooms at the surface contribute a large volume of dead organic matter to the quieter, deeper waters. The microbial consumption of this material as it falls through the water column depletes dissolved oxygen, leading first to an oxygen-minimum layer, and eventually to a completely anoxic basin, at least below the thermocline or halocline. Such stagnant basins or pools are well known from marginal, semi-enclosed seas, such as the Black Sea, Eastern Mediterranean, or Gulf of California, but are less easily generated in a large open ocean. When and where they do occur, the destruction of organic matter is much reduced so that it accumulates on the seafloor as carbon-rich sediment. This is a significant sink within the global carbon cycle, as well as a future source for generating oil.

Over the past million years there have been at least ten episodes of stagnation in the Eastern Mediterranean, each yielding a thin layer of black carbon-rich sediment (black shale) across the whole basin. During similar periods of the more distant past, known as ocean anoxic events, much larger parts of the global ocean appear to have reached such low levels of oxygen that widespread black shales have accumulated. But the ocean is inherently unstable and such stagnation cannot persist for long before the ocean waters are overturned and oxygen once more introduced into the deep. Layers of black shale are therefore interbedded with layers of pale-coloured, carbon-poor sediment.

Chapter 5
Dynamic ocean

The relentless pounding of waves against a rock-strewn shore
and rhythm of tides that sweep back and forth across mud flats
and broad sandy beaches are the very heartbeat of the ocean.
They have been constant and regular since the oceans first formed
4 billion years ago. For mariners, fishermen, leisure-time sailors, or
passengers, the endless motion of the sea surface is an unpredictable
force to be reckoned with. There are feared and notorious parts of
the ocean that any sailor would prefer to avoid, even in today's world
of supreme technology. Ocean currents are many thousand times
more powerful than any river on land. Those in the deep ocean are
silent and unseen. Here, there are waterfalls without sound, rivers
without banks, and storms that rage unnoticed for weeks at a time.
This chapter explains the fuzzy logic of ocean dynamics.

It is the dynamics of the sea surface that causes so many people to
suffer motion sickness at sea. I am generally little affected, which
is one of the reasons I so much enjoy ocean-going expeditions.
But everyone has their natural limit, as I found to my cost on one
particular voyage to the eastern Pacific Ocean. Mine is prolonged
Gale Force 9 on a narrow-hulled ex-navy ship when our survey
course took us oblique to the wind and 10-metre waves—the yaw
was horrendous, so that most of the crew and all the scientists
suffered.

Our findings, on the other hand, were very exciting. We were surveying an area of the seafloor adjacent to the Monterey deep-sea channel, where turbidity currents had overspilled the channel bank as they rounded a meander and deposited a field of gigantic sediment waves—just like ripples in the sand, except that each had a wavelength of around 2 kilometres and a height of 40–50 metres. I published a new theory for their origin from turbidity currents, with a physical oceanographer colleague, Tony Bowen, from Dalhousie University. Such fields of sediment waves are now well known from many parts of the ocean floor and several different theories for their origin exist, testament to the range of currents and processes that operate in the deep sea (Figure 10, Chapter 3).

Energy in motion

Almost all the surface waves we observe are generated by wind stress, acting either locally or far out to sea. Although the wave crests *appear* to move forwards with the wind, this does not occur. Mechanical energy, created by the original disturbance that caused the wave, travels through the ocean at the speed of the wave, whereas water does not. Individual molecules of water simply move back and forth, up and down, in a generally circular motion. This is shown quite clearly by a resting seagull or floating driftwood that bobs up and down as the wave energy passes, but moves only very slightly forwards with each wave (Figure 15).

The greater the wind force, the bigger the wave, the more energy stored within its bulk, and the more energy released when it eventually breaks. The amount of energy is enormous. Over long periods of time, whole coastlines retreat before the pounding waves—cliffs topple, rocks are worn to pebbles, pebbles to sand, and so on. Individual storm waves can exert instantaneous pressures of up to 30,000 kilograms (3 tonnes)

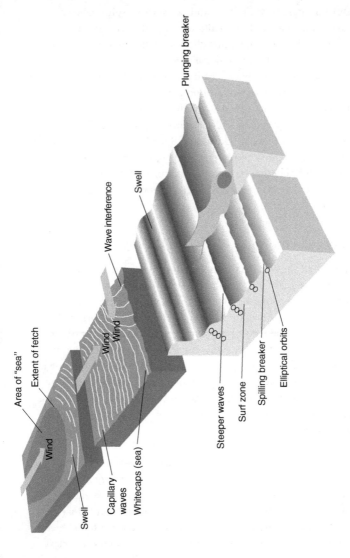

15. Wave generation and types.

Plunging breaker

Wave interference

Swell

Area of "sea"

Extent of fetch

Wind

Wind

Wind

Swell

Capillary waves

Whitecaps (sea)

Steeper waves

Surf zone

Spilling breaker

Elliptical orbits

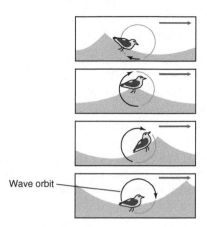

Wave orbit

15. Continued.

per square metre. Concrete piers can be removed by a single massive wave; whole houses and villages can be washed away in the course of a major storm. In open waters, even large ships can be tipped over and under, or tossed head over heels and snapped like matchsticks.

The rate at which energy is transferred across the ocean is the same as the velocity of the wave. This can be determined from a simple physical relationship in which:

$$\text{Wave velocity} = \frac{\text{Wavelength (distance between crests)}}{\text{Wave period (time interval between waves)}}$$

On average, successive waves are between 60 and 150 metres apart, and the time period between successive wave crests is 10–15 seconds. It follows from the above relationship that waves typically travel at speeds of 30–40 kilometres per hour, and that waves with a greater wavelength will travel faster than those with a shorter wavelength.

Sea, swell, and surf

Wind-generated waves form on the sea surface by the transfer of energy from the wind to water. The smallest waves formed by a gentle puff of wind that is barely sufficient to break the surface tension are tiny diamond-shaped ripples known as cats' paws. With increasing wind speed and duration over which the wind blows, the wave height, period, and length all increase. The distance over which the wind blows is known as fetch, and is critical in influencing the growth of waves—the greater the area of ocean over which a storm blows, then the larger and more powerful the waves generated.

The three stages in wave development are known as sea, swell, and surf. The state of random choppiness, building to larger but still irregular waves with no systematic pattern, is known simply as sea. As these waves leave the region of fetch where they were generated, the longer waves outpace the shorter ones because their velocity is greater. Gradually, they fall in with other waves travelling at similar speed. Where different waves are in phase they reinforce each other and where out of phase they are reduced in size. Eventually, a regular pattern of high and low waves is developed that remains constant as it travels out across the ocean. This is known as the swell.

The ocean is highly efficient at transmitting energy. Water offers so little resistance to the small orbital motion of water particles in waves that individual wave trains may continue for thousands of kilometres. They decrease in height a little as they lose some energy, but the same regular pattern remains. Major storm waves generated in the Southern Ocean off Antarctica can take nearly a week crossing the Pacific before they break along the shores of Hawaii. Those that miss the islands travel on for another three or four days before they wash ashore on the remote beaches of Alaska. Their size has diminished but the original pattern of swell that first left the Southern Ocean remains intact.

When the wave train encounters shallow water—say 50 metres for a 100-metre wavelength—the waves first feel the bottom and begin to slow down in response to frictional resistance. Wavelength decreases, the crests bunch closer together, and wave height increases until the wave becomes unstable and topples forwards as surf. Violent plunging waves, with that characteristic tube of air that surfers love to ride, form when large waves approach a steeply sloping bottom. A more gently sloping bottom generates a milder spilling wave.

Very often, waves approach obliquely to the coast and set up a significant transfer of water and sediment along the shoreline. The longshore currents so developed can be very powerful, removing beach sand and building out spits and bars across the mouths of estuaries. The build-up of energy in longshore currents becomes sufficient to overcome the power of incoming waves at an irregular spacing along the coast. This leads to the development of strong offshore-directed rip-currents, which are especially dangerous to swimmers, and instrumental in the seaward transport of sediment.

Catastrophic waves

Rogue waves cannot be predicted, they strike erratically, often in the middle of the ocean, and then disappear without trace. They occur when two or more storm wave crests merge, or when a series of storm waves encounters an opposing current. When the current speed exceeds 4 knots, opposing storm waves can whip upwards to four times their original height, hurtle forwards, and break at sea under their own instability. Examples are well known from the Agulhas Current off the South African Wild Coast, where shipwrecks litter the seabed, as well as from the Kuroshio Current off Japan and the Gulf Stream in the North Atlantic. Particular climatic conditions that lead to higher than normal storm waves include hurricanes, typhoons, tornadoes, thunderstorms, downbursts, and sudden pressure drops.

Tsunamis are unusual waves of extreme wavelength (up to 200 kilometres) that travel radially away from their point of origin at speeds up to 800 kilometres per hour—similar to that of a long-haul passenger jet. They are not due to wind stress but are caused by the sudden vertical movement of the seabed along faults, due to subsea earthquakes or giant submarine slides, as well as by sudden displacement of the ocean surface. This can result from a major landslide or debris avalanche moving directly from land into the sea, from huge icebergs calving and falling from the edge of a glacier or ice-cliff, and from major explosive volcanic flows into water. In each case, it is the enormous volume of water displaced that gives them such power and velocity.

The worst tsunami of recent times occurred just after midnight on 26 December 2004. It was the direct result of a seafloor earthquake off Indonesia—the third largest ever recorded (9.1–9.3 on the Richter scale), releasing energy equivalent to 23,000 Hiroshima-sized atomic bombs, and causing the whole Earth to wobble by as much as 1 centimetre. The tsunami travelled clear across the Indian Ocean, reaching Sri Lanka in 2.5 hours and South Africa in 11 hours. In the open ocean the wave height was around 50 centimetres and passed completely unnoticed by most of the ships, large and small, in its path. But as the tsunami approached land the velocity slowed and the wave height increased dramatically, reaching up to 30 metres in a very short period of time on the Indonesian island of Sumatra.

Rhythm of the tides

Tides are created by the gravitational pull of the Moon and the Sun and by a centrifugal force due to the rotation of the Earth. Because the ocean waters are mobile with respect to the solid Earth, lunar gravity is able to pull the water towards it, creating a very slight bulge on the side of the Earth nearest the Moon.

On the opposite side, centrifugal force pulls water away from the Moon, creating a second bulge. These opposing bulges are the high tides. In between, where the water has pulled away, are the areas of low tide (Figure 16).

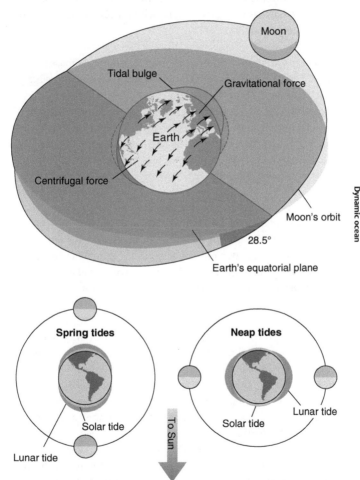

16. **Tides and how they work.**

As the Earth spins on its axis every 24 hours, so each point should pass through each of these opposing bulges, thus experiencing two high tides and two low tides in a daily cycle. Indeed, this is very nearly true in most of the world. But friction slows the movement of water a little, while the Moon's orbit around the Earth is not quite in phase with the Earth's spin, gaining by 50 minutes each day.

The Sun is 400 times further away from Earth, yet its mass is about 27 million times that of the Moon. In effect, this translates to the fact that the Sun exerts a gravitational pull on the ocean waters that is about 40 per cent as strong as the lunar influence. The tidal bulges caused by the Sun's gravity and opposing centrifugal force, move around the Earth with a quite different periodicity from those due to the Moon—a solar year rather than the lunar month. When the Earth, Moon, and Sun are aligned, at times of full moon or new moon, then the tidal bulges are in phase leading to extra-high and extra-low tides. These twice-monthly extremes of tidal range are known as spring tides, because they seem to well up like a spring (nothing to do with *spring* as a season). When the Moon and the Sun are at right angles to each other with respect to the Earth, then the gravitational attractions are opposed and the tidal range is least. These are known as neap tides.

Much of the global ocean experiences semidiurnal (twice daily) tides that are approximately equal in height. Some of the western Pacific, however, experiences only one high and one low tide per day (diurnal). Tidal ranges across most of the world are between 1 and 3 metres, although in semi-enclosed seas, such as the Mediterranean, and the Black and Red Seas, they are almost imperceptible. Semi-restricted bays or funnel-shaped estuaries, on the other hand, serve to greatly exaggerate tidal range—the maximum recorded being 16 metres in the Bay of Fundy, eastern Canada, and over 12 metres along the Bristol Channel and Severn Estuary, south-western England.

Great surface currents

Wind is the principal force that drives surface currents, but the pattern of circulation results from a more complex interaction of wind drag, pressure gradients, and Coriolis deflection. Wind drag is a very inefficient process by which the momentum of moving air molecules is transmitted to water molecules at the ocean surface setting them in motion. The speed of water molecules (the current), initially in the direction of the wind, is only about 3–4 per cent of the wind speed. This means that a wind blowing constantly over a period of time at 50 kilometres per hour will produce a water current of about 1 knot (2 kilometres per hour).

The second principal force influencing how the pattern of surface currents develops is that caused by seawater being piled up into mounds. This may seem counter-intuitive but, far from being a flat surface once we have removed the local ups and downs of waves, the ocean surface is actually warped into broad mounds and depressions. Converging currents and persistent onshore winds tend to pile water up faster than it can flow away, while diverging currents result in a drawdown of water and the creation of lows.

The third factor involved is the Coriolis force. This force is the result of planetary rotation and is felt most keenly by all moving objects (water, wind, aeroplanes, etc.) that are not rigidly attached to the Earth's surface. Moving currents of water and of air are affected in exactly the same way by the Coriolis force, veering to the right in the northern hemisphere and to the left in the southern hemisphere (Figure 17).

Gyres, loops, and eddies

Working in close consort, these various forces control the surface movement of ocean water. Although the movement of wind may seem random, changing from one day to the next, surface winds actually blow in a very regular pattern on a planetary scale.

Oceans

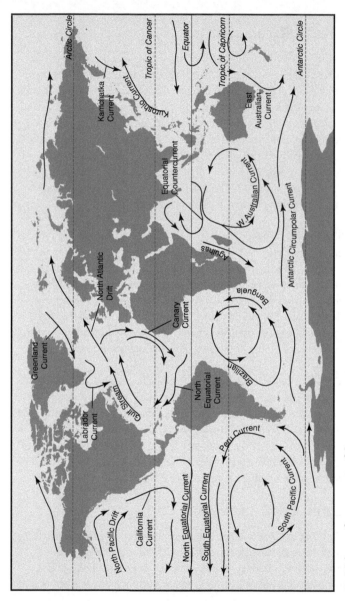

17. **Major surface currents of the oceans.**

The subtropics are known for the trade winds with their strong easterly component, and the mid-latitudes for persistent westerlies. Wind drag by such large-scale wind systems sets the ocean waters in motion. The trade winds produce a pair of equatorial currents moving to the west in each ocean, while the westerlies drive a belt of currents that flow to the east at mid-latitudes in both hemispheres. (Don't be confused—sailors and oceanographers always call a current flowing eastwards an easterly current, while landlubbers and meteorologists call a wind blowing eastwards a westerly!) Deflection by the Coriolis force and ultimately by the position of the continents creates very large oval-shaped gyres in each ocean.

There is a mirror image of flow about the equator yielding opposing gyres in the northern and southern parts of the oceans. The Antarctic Circumpolar Current is not interrupted by any landmass and so flows continuously around Antarctica, driven by the constant action of the Roaring Forties and Fifties.

As the Coriolis effect continues to deflect the wind-generated currents, so it causes a net transfer of water at right angles to the direction of the wind. This in turn creates a gigantic mound of water within the central region of each ocean gyre. Water tries to flow back down the pressure gradient so produced, and is further deflected by the Coriolis force as it moves. This leads to generation of a steady-state current known as a geostrophic current. The water at the centre of these geostrophic gyres is almost stationary as the main current flows around the outer ring.

In addition, water in the oceans tends to pile up against the continental landmasses along their western margins. Here the geostrophic flows become confined and thereby intensified with respect to the broader, weaker return flows on the eastern margins. Such intensification further produces flow instabilities so that western boundary currents, in particular, are seen to weave and meander along their course in snake-like fashion. Tight meander loops form, sometimes pinching off completely to form large swirling eddies or rings that spin away from the main flow mixing into the adjacent seas.

Wind drag moves water molecules and the motion of these in turn creates a fluid drag on those just below the surface, and so on downwards until the wind energy is fully dissipated. Coriolis deflection causes each successive layer to deviate very slightly from that above. The result is a spiralling flow pattern of decreasing velocity downwards, like a whirlpool with an almost imperceptibly slow spin, known as the Ekman spiral, after the Scandinavian physicist who first explained the phenomenon. The net transfer of water at right angles to the wind direction is called Ekman transport, and this is responsible for another extremely important effect—that of coastal upwelling. In certain parts of the oceans, there is a net transfer of surface water offshore, which is replaced by upwelling of water from below. These deep-sourced waters are rich in nutrients, derived from the dissolution and recycling of slowly sinking planktonic organisms, and so support a great profusion of marine life.

Global conveyor belt

Entirely hidden from view in the deep-ocean basins and involving 90 per cent of ocean waters, are powerful and omnipresent, but very slow-moving currents. Linked with the great surface gyres, these currents form a vast oceanic network of circulating water that transfers energy, nutrients, and sediments around the world. Driven not by atmospheric winds but by density differences linked to water temperature and salinity, these bottom currents are part of the global thermohaline circulation system (Figure 18, Table 4).

The pattern of deep thermohaline circulation in the oceans starts at the surface, mainly at very high latitudes. These are the cold-water kitchens of bottom-water production. During the long polar winters when sunlight is absent or minimal for 24 hours a

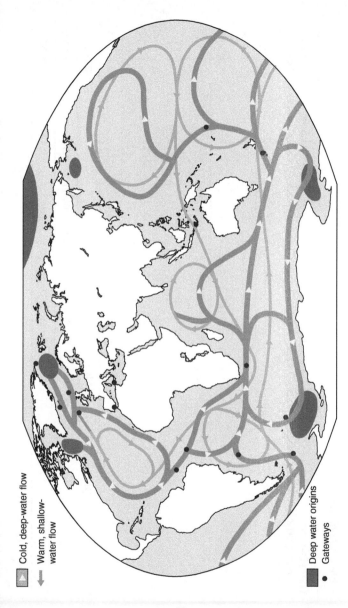

Cold, deep-water flow

Warm, shallow-water flow

Deep water origins

Gateways

18. Deepwater currents and thermohaline circulation.

Table 4 Water mass properties

Global water mass properties	Types (depth range)	Water mass	Temperature (°C)	Salinity (‰)
	CENTRAL (0–1 km)	Pacific Central Water	7–20	34.1–36.2
		Atlantic Central Water	4–20	34.3–36.8
		South Indian Central Water	6–16	34.5–35.6
	INTERMEDIATE (1–2 km)	North Pacific Intermediate Water	4–10	34.0–34.5
		Red Sea Intermediate Water	23	40.0
		Mediterranean Intermediate Water	6–11.9	35.3–36.5
		Arctic Intermediate Water	0–2	34.9
		Antarctic Intermediate Water	2.2–5	33.8–34.6
	DEEP/BOTTOM (over 2 km)	Common Water	0.6–9	33.5–34.7
		North Atlantic Deep/Bottom Water	2.5–4	34.9–35.0
		Antarctic Deep/Bottom Water	−0.4–4	34.6–35.0

day, seawater becomes extremely cold and hence much denser. At the same time, sea ice expands out from the continents and further increases the density of the water beneath. This is because the freezing process incorporates mostly freshwater, making the underlying water more saline. Cold, saline, dense water sinks rapidly and spreads out across the ocean floors, moving very slowly towards lower latitudes. At the ocean surface, currents flowing from the equator to the poles replace the waters that have sunk with warmer water, which in its turn will cool and sink. The cycle is apparently without end and has become known as the global ocean conveyor belt.

The principal cold-water kitchen off Antarctica is the Weddell Sea, where the coldest and densest of all water masses is produced. Antarctic Bottom Water, as it is called, cascades down the steep continental slope off Antarctica and spreads out across the floor of the ocean. It flows north into the Atlantic, crosses the equator, and finally mixes upwards, losing its identity somewhere east of Newfoundland. Slightly less dense waters are formed in a broad region of the Southern Ocean, before they too plunge beneath the warmer subpolar waters at the Antarctic Convergence zone around 60°S.

Cold bottom waters that form beneath sea ice in the Arctic Ocean are mainly trapped there by the high sills that surround the basin. The principal cold-water kitchen in the northern hemisphere, therefore, is the winter cooling of surface waters in the Norwegian and Greenland Seas. As they spill across the Denmark Strait and through the Faeroe-Shetland Channel, they mix with a small amount generated in the Labrador Sea to form North Atlantic Deep Water. This flows southwards until it ultimately blends with Antarctic Bottom Water to form a water mass known simply as Common Water. It is this mixture that dominates the deep Indian Ocean and much of the Pacific as well.

Submarine gateways and waterfalls

As noted in Chapter 3, the oceans are compartmentalized into abyssal basins separated by submarine mountain ranges and plateaus. Along the length of the global conveyor belt, bottom waters pile up behind such topographic barriers until they reach the spill point. Typically, this occurs as a narrow valley that cuts across the barrier—known as an ocean gateway. As the huge weight of dense water funnels through the narrow gateway, it is severely restricted in width and so accelerates. The bottom currents through such deep-ocean passageways can be highly abrasive, scouring away loose sediment and eroding bare rock. Bottom current velocities of 1–2 metres per second (4–8 kilometres per hour) are quite common—this is strong and fast for the otherwise peaceful world of the deep sea.

As the narrow high-velocity bottom current enters the adjacent basin, the dense water spreads out and cascades downslope. We call these features submarine waterfalls because of their immense height and power, although with such gentle slopes in reality they are more akin to the cataracts or rapids of terrestrial rivers. Whatever the slope, their scale is dramatic. Cold Antarctic Bottom Water, only a fraction of a degree above freezing point (at about 0.2°C) piles up behind the Rio Grande Rise in the South Atlantic and then cascades into the Brazilian Basin—a drop of over 1000 metres. Still more impressive is the submarine waterfall located beneath the Denmark Strait in the North Atlantic. Here, 5 million cubic metres of water cascades downslope into the North Atlantic basin every second, generating giant eddies and turbulent whirlpools, and drops a vertical distance of over 3.5 kilometres. The Denmark Strait waterfall dwarfs any similar features on land. The highest waterfall is the Angel Falls in Venezuela, which drops a mere 1 kilometre. The Cuaira Falls on the Paraguay–Brazil border have the largest average flow rate, 13,000 cubic metres per second, which is 4000 times less than their submarine counterpart.

Chapter 6
Ocean–climate nexus

The oceans and atmosphere are intricately coupled. Together they control and express both the daily drama of Earth's weather systems and the long-term changes in planetary climate. Winds drive the currents that redistribute heat across the face of the globe—a transfer of energy that is staggering in its enormity, and essential to the maintenance of a habitable world. But the climate is no more constant than its ocean regulator is simple. The natural state for Earth during most of its history has been one of warm (greenhouse) conditions. These have been sporadically interrupted by cold (icehouse) conditions with a long-period cyclicity that we can recognize, but not yet fully understand. For the past 3 million years, Earth has been in the grips of one of these icehouse periods, during which glacial and interglacial periods have alternated with regularity. The human impact has been to influence these natural cycles of change dramatically, enhancing the greenhouse effect and causing sea level to rise. This chapter aims to cut through to the essence of this very complex and emotive subject.

Recent climate change

The international consensus is that the Earth is warming rapidly under a thick cloak of greenhouse gases, and that humankind is now the principal contributor to this atmospheric build-up. Certainly, at the start of the 21st century, the world appears to be a warmer place than at any time in the past 700–800 years. Since records began in 1880, the global average surface temperature has increased by about 0.8°C, and is predicted to rise by a further 1.4°C to 5.8°C over the next century. Fifteen of the warmest years on record have been in the 21st century; each decade is warmer on average than the previous one.

As a direct result of global warming, average sea level has risen by 10–20 centimetres through the 20th century as polar ice sheets begin to thin. It is expected to rise some 25–100 centimetres by 2100, as melting continues and because seawater expands slightly as it warms. In a worst-case scenario, a great many island nations would vanish altogether, along with a number of major coastal cities—Shanghai, Venice, and New Orleans, for example. Some 15 per cent of Bangladesh would be submerged, displacing 15 million people. Coastal erosion and widespread flooding would increase and so threaten homes, farmland, and natural habitats in low-lying areas everywhere. This is not sensationalism; it is simply the inevitable effect of sea-level rise.

Extremes of weather are already becoming more common and this trend would continue as temperatures increase further. In middle to high latitudes of the northern hemisphere, cloud cover has increased and heavy rains are more frequent. El Niño events have become more marked over the last forty years. Tropical cyclones and temperate-area storms seem likely to intensify further, bringing torrential rains to some places and drought to others. Some of these changes are positive—harsh climes becoming more equable, greening of desert areas, and longer growing seasons in high latitudes—but the regional patterns of change, their influence on global productivity, and the long-term ocean–climate response are still very poorly understood.

What we do understand is that incident sunlight warms the planet. Some of the solar energy is absorbed directly, but much is emitted back into space as infrared radiation. Certain 'greenhouse' gases in the atmosphere—notably carbon dioxide, water vapour, and methane—absorb part of this infrared energy, trapping it in the atmosphere and so making Earth a far warmer and more habitable planet than it would otherwise be. This is known as the *greenhouse effect.*

Stirring the atmosphere

The envelope of gases forming the atmosphere extends 2400 kilometres above the surface of the Earth and comprises four principal layers. The troposphere is the lowest and thinnest layer (only some 10–12 kilometres thick) and contains 75 per cent of the mass. It is made up of 78 per cent nitrogen and 21 per cent oxygen, as well as a host of other gases in minute quantities (Table 5). It also contains virtually all the moisture and hence all the clouds, snow, rain, and winds—very much the active part of the atmospheric heat engine.

Three planetary-scale spiralling circuits of wind dominate each hemisphere (Figure 19). They are set in motion by a differential

Table 5 Principal gases in the atmosphere

Principal gas	Concentration (%)
Nitrogen (N)	78.08
Oxygen (O)	20.95
Argon (Ar)	0.93
Water vapour (H_2O)	0.1–1.0 (variable)
Carbon dioxide (CO_2)	0.04
Methane (CH_4)	0.000172
Nitric oxide (N_2O)	0.000031
Ozone (O_3)	0.000005 (variable)

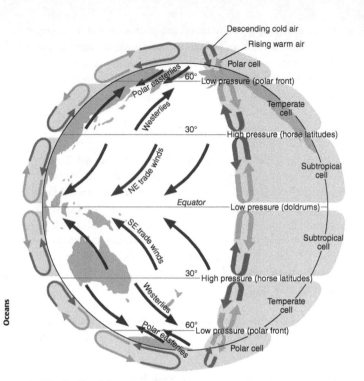

19. Wind cells of the atmosphere.

influx of solar energy, directed by Earth's rotation and deflected by
the Coriolis force. The two subtropical cells straddling the equator
experience persistent trade winds. Between the north-east (NE)
and south-east (SE) trade winds lie the doldrums, with constant
hot temperatures, little wind movement, and high humidity.
The temperate cells at mid-latitudes are buffeted alternately by
mild westerlies, sweeping warm air away from the subtropical
high-pressure zone towards the polar front, and harsh polar
easterlies that form in the polar cells. Cold frontal depressions or
cyclones are generated at the fluctuating interface between these
two opposing wind systems. These lead to frequent storms, which

become stronger during winter months when the pole–equator temperature gradient increases, driving the heat engine that much faster.

Above the spiral wind cells of the troposphere are narrow, high-velocity jet streams, between 10 and 15 kilometres high, with wind speeds exceeding 160 kilometres per hour. Even at these speeds, the jet streams follow sinuous paths, diving and rising as they flow, and help to create and direct cyclones and anticyclones below. During winter, speeds increase to as much as 500 kilometres per hour, the jet streams are straighter, and the storm winds are stronger.

The distribution of oceans and continents and the height of mountains across the continental interior have an added effect on the winds. This is the seasonal reversal of winds, known as monsoonal circulation. It is caused by the unequal heating of the low-latitude continents and tropical oceans as the Sun swings back and forth between the two hemispheres during its annual cycle. During the northern hemisphere winter, while the Sun is overhead in the south, the continental landmass of Eurasia and North Africa cools dramatically. With little moisture to pick up and high pressures due to the cold heavy air, the north-easterly winds that blow outwards from the continental interior are cool and dry. As the winds blow across the ocean, they gather moisture and deliver torrential rains to South East Asia, northern Australia, central Africa, and South America—the southern hemisphere monsoons.

A reversal of climatic conditions during the northern hemisphere summer leads to continental heating, especially over the Himalayas and Tibetan Plateau, and rising air masses. Moisture-laden winds are drawn in from the oceans and the northern hemisphere monsoon rains pour down from May through September each year. They are particularly heavy in South and South East Asia.

Ocean regulator

The control exerted by the oceans is an integral and essential part of the global climate system. The immense amounts of heat and moisture stored in the oceans act as a giant flywheel to the climate system, both moderating change and also prolonging it once change commences. The oceans are one of the principal long-term stores on Earth for carbon and carbon dioxide and so act to regulate greenhouse gases in the atmosphere, although with an efficacy we cannot yet predict.

A crucial piece of the climate jigsaw, therefore, is the way in which greenhouse gases are transferred between air and sea. The oceans are like a gigantic sponge holding fifty times more carbon dioxide than the atmosphere, and they are thought to absorb between 30 per cent and 40 per cent of the carbon dioxide produced by human activity. But the sea surface acts as a two-way control valve for gas transfer, which opens and closes in response to two key properties—gas concentration and ocean stirring.

First, the difference in gas concentration between the air and sea controls the direction and rate of gas exchange. Gas concentration in water depends on temperature—cold water dissolves more carbon dioxide than warm water, and on biological processes—such as photosynthesis and respiration by microscopic plants, animals, and bacteria that make up the plankton. These transfer processes affect all gases, including dimethylsulfide (DMS), which is produced in abundance during phytoplankton blooms. When DMS escapes into the atmosphere it aids in the formation of clouds and hence changes the reflective properties of the atmosphere.

Second, the strength of the ocean-stirring process, caused by wind and foaming waves, affects the ease with which gases are absorbed at the surface. More gas is absorbed during stormy weather and, once dissolved, is quickly mixed downwards by water turbulence.

Both physical and biological pumps then act to remove carbon dioxide from the ocean surface, so that it remains out of contact with the atmosphere for long periods. The physical pump works because cold water, charged with dissolved carbon dioxide, is heavier than warm water and so sinks to the seafloor. The biological pump works through the uptake of carbon dioxide by planktonic organisms. After death, this carbon-rich matter sinks through the water column and is progressively broken down by bacteria and chemical corrosion as it falls. Bottom-dwelling animals consume much of what eventually reaches the seafloor. Some of the carbon is returned to the water, while another part is buried with the sediment.

The transfer of heat, moisture, and other gases between the ocean and atmosphere drives small-scale oscillations in climate. The El Niño Southern Oscillation (ENSO) is the best known, causing 3–7-year climate cycles driven by the interaction of sea-surface temperature and trade winds along the equatorial Pacific. The effects are worldwide in their impact through a process of atmospheric teleconnection—causing floods in Europe and North America, monsoon failure and severe drought in India, South East Asia, and Australia, as well as decimation of the anchovy fishing industry off Peru.

Tropical cyclones

The largest, most violent storms on Earth are generated at sea. These are known by various names around the world—in the Indian Ocean they are called cyclones, meaning coiled serpents; in the western Pacific they are typhoons, after the Chinese word Tai Fung for great wind; and in the Atlantic they are hurricanes, after Hunraken, the Mayan god of winds. The energy released by a single tropical cyclone in one day would be enough to power the entire industrial production of the United States for one year.

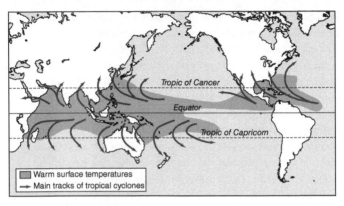

20. Tropical cyclone formation.

Tropical cyclones are the direct result of overheating of the oceans (Figure 20). They develop over warm seas at tropical latitudes when sea-surface temperatures exceed 27°C. As vast amounts of water evaporate from the overheated ocean surface, the hot moist air rises and then condenses as it cools to form billowing cumulonimbus clouds. Air rushes inwards across the sea surface to fill the void, evaporating more water as it passes, causing more clouds to form. Spiralling bands of thunderstorms begin to rotate anticlockwise in the northern hemisphere (clockwise in the southern hemisphere), around a calm clear eye of the storm. As they grow, so they become self-sustaining, sapping heat energy from the ocean and driving the spinning winds ever faster, with wind speeds typically exceeding 160 kilometres per hour and locally gusting to over 350 kilometres per hour. Cyclones extend over a circular area from 500 to 1500 kilometres in diameter.

The role of ozone

Above the troposphere lies the stratosphere, which is 35–40 kilometres thick. Tiny amounts of water vapour sometimes escape from the troposphere to form thin, colourful nacreous clouds. The stratosphere also contains large amounts of very highly dispersed ozone gas—a type of oxygen made up of three oxygen atoms rather than the two atoms forming normal oxygen. It is this ozone layer that protects all life on Earth from the harmful effects of the Sun's ultraviolet rays. It acts as an invisible sunscreen every bit as essential for our survival as the oxygen we breathe.

Since oxygen first appeared in the atmosphere 2.5 billion years ago, building up to its present level about 650 million years ago, there has been an ozone layer above kept in a state of dynamic balance by a complex array of chemical reactions. However, a dramatic rise in pollution since the 1960s, mainly involving chlorine gas from chlorofluorocarbons (CFCs) used in aerosols, refrigerators, and air conditioning systems, has upset this chemical balance and resulted in the destruction of ozone.

Today, there is a marked thinning of the ozone layer in an area about the size of North America, which appears over Antarctica each austral spring. A smaller area of thinning also appears over the Arctic. These regions of thinning are commonly referred to as *ozone holes*. The consequences of such ozone loss are very serious. A 10 per cent reduction in ozone will result in a 20 per cent increase in the amount of ultraviolet radiation reaching the Earth's surface—this will inevitably lead to an increase in skin cancer and eye cataracts, as well as a general suppression of the immune system. Penetration of ultraviolet radiation to the ocean surface will damage plankton, with obvious repercussions higher up the marine food chain.

The role of ozone in climate is less clear. Ozone in the stratosphere has a cooling effect, so that its reduction will tend to lessen this. There is also ozone in the troposphere, mainly as a by-product of traffic pollution, where it makes a very small (<1 per cent) contribution to global warming as a greenhouse gas. In fact, CFCs make a larger contribution to warming. Fortunately, however, the world has now agreed to ban CFCs and further seek to reduce chlorine emissions. If this is achieved and maintained, then the ozone levels will fully regenerate with time.

Past climate change

Earth's climate has not always been as it is today (Figures 21, 22). There are times when the Earth has experienced warm, humid conditions with little temperature difference between the equator and the poles. About 100 million years ago, for example, palm trees and crocodiles lived as far north as 80°N—the equivalent of Arctic Canada or northern Greenland today. At other times a far colder climate has gripped the planet, so that the area covered by ice sheets, glaciers, and frozen tundra greatly expanded. Past climate changes had entirely natural causes.

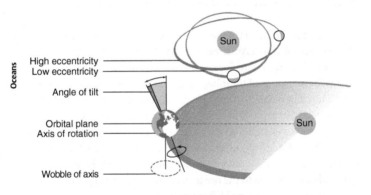

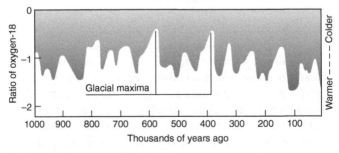

21. Oxygen isotopes and climate cycles.

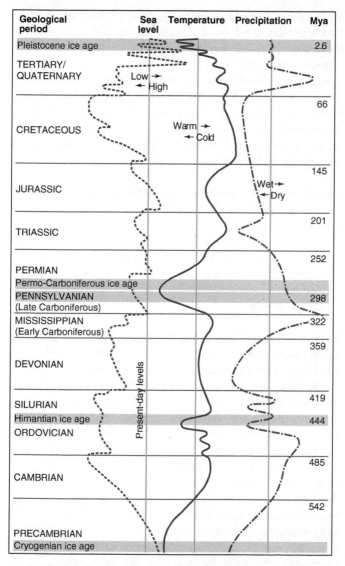

22. Sea-level and temperature chart through time.

Most of the geological past has enjoyed warm conditions. These have been interrupted at irregular intervals by cold and glacial climates of altogether shorter duration—the greenhouse and icehouse conditions, respectively. We do not yet have a full scientific explanation for the causes and timing of such large-scale climatic fluctuations. We do know that the solar influx of heat, and moderation by the ocean–atmosphere couple, are two of the key drivers involved. We also know that the size and distribution of oceans, continents, and ice caps are important in regulating the amount of solar heat that is absorbed or reflected.

There is some evidence that the rocks of the Gowganda series in Canada, deposited more than 2 billion years ago, represent the very first icehouse world. Since that time, we can be more confident in identifying six principal episodes—very approximately at 950, 750, 650, 450, and 300 million years ago, and the last beginning around 3 million years ago. We are still in the grip of this last icehouse state, although in one of its relatively brief interglacial phases.

It arrived quite suddenly and with dramatic effect. The global climate began to fluctuate wildly, but with an overall cooling trend. Winds grew stronger, rain and snow increased, and the inevitable cloud build-up served to reflect more of the Sun's heat away from the Earth's surface. World temperatures dropped by up to 8°C and the sea-surface temperatures by as much as 6°C. Polar ice caps spread from the Antarctic and Greenland, as huge floating ice shelves around Antarctica and over the Arctic Ocean and the Norwegian, Labrador, Hudson, Barents, and Bering Seas. Ice sheets formed in mountain and highland regions and then spread into temperate lowland areas of the northern hemisphere, covering half of the land area of North America and a quarter of Eurasia with ice 2–3 kilometres thick. At times almost one-third of the continental land mass was covered by ice, with thicknesses at the heart of polar ice caps reaching 6–7 kilometres.

During the past 3 million years of icehouse conditions, global climate has varied in regular cyclic fashion due to several forcing factors. The most important of these are changes in Earth's orbit around the Sun, which affect climate in periods of about 22,000, 40,000, and 100,000 years—known as Milankovitch cycles. There have been between ten and eleven of the longer period cycles in the last 1 million years, pushing the Earth from glacial to interglacial conditions. The most recent interglacial episode began 11,000 years ago. Smaller scale wobbles and oscillations in this regular pattern are also evident and known as sub-Milankovitch cycles. There have been relatively warm periods peaking at 6500, 4500, 2000, and 1000 years ago, separated by colder periods, for example the Little Ice Age between 1645 and 1715 CE. These small-scale variations are due to a combination of oceanic forcing and irregularities in solar insolation.

An important discovery of the 1970s, pioneered by Sir Nick Shackleton and colleagues at Cambridge University, has allowed us to build up an ever more refined record of past climate change. They showed that the two principal forms of oxygen (known as oxygen-18 and oxygen-16 isotopes) occur in very slightly different proportions in the water molecules of ice as opposed to normal seawater. When large amounts of water are locked up in the polar ice caps, then the isotopic composition of seawater is preferentially enriched in the heavy oxygen-18 isotope. This is directly transferred to the chemical make-up of fossil shell material secreted by marine organisms, which therefore records the isotopic proportions of the oceans in which they lived.

By carefully picking microfossils from sediment cores and analysing their isotopic chemistry, we have slowly built up a complete record of seawater variation through the past 140 million years. This marine oxygen isotope stratigraphy is the best *proxy* we now have for past climate change and sea-level variation and provides the worldwide standard. Other materials that have oxygen in their make-up can also be used—frozen water from ice cores drilled into the polar ice caps, or stalagmites precipitated in underground caves, for example. All these records compare very closely with one another and with the climate variations predicted by the theory of Milankovitch cycles (Figure 22).

One of my long-term scientific interests has been the study of *contourites*—sediments deposited in the deep sea by bottom currents. In 2012, we cored and recovered over 4500 metres of these sediments from the Atlantic continental margin south of Iberia. To my great relief, our findings confirmed the contourite model I had proposed some thirty years previously. We were also able to use other proxies, in addition to oxygen isotopes, to infer climate change. For example, changes in mean grain size of the contourites could be linked to climate-driven fluctuation in bottom current velocity, and systematic colour changes reflected a climate-controlled influx of sedimentary material across the continental margin. Our understanding of the climate–ocean nexus becomes ever more refined.

Sea-level rise and fall

Sea level has varied in the past in close consort with climate change (Figure 22). Around twenty-five thousand years ago, at the height of the last Ice Age, the global sea level was 120 metres *lower* than today. Huge tracts of the continental shelves that rim today's landmasses were exposed. Land bridges opened up across the world—Alaska to Siberia, the British Isles to continental Europe, and many others. Further back in time, 80 million years ago, the sea level was around 250–350 metres *higher* than today, so that 82 per cent of the planet was ocean and only 18 per cent remained as dry land. Such changes have been the norm throughout geological history and entirely the result of natural causes.

We understand these natural causes of sea-level change, although we do not fully understand their triggers. The four principal causes are:

(1) Displacement of seawater upwards by the addition of new ocean crust. This pushes the seafloor upwards due to the growth of mid-ocean ridges, oceanic hot spots and submarine volcanic plateaus. When spreading rates are high and the outpouring of volcanic lava greatest, then sea level rises. When rates are slow and volcanic activity is less, the sea level falls. Both these effects are global in their extent.

(2) Mountain uplift and crustal subsidence. This is the natural result of plate tectonic movement and creates local changes in sea level. In this case, sea-level rise and fall is relative and regional. Large-scale mountain uplift can also induce climate change and hence become a global driver.

(3) Changes in the distribution of continents and oceans due to plate tectonic movement. When tectonic drift moves a continental landmass over the polar region then it can allow the accumulation of snow and ice on land and hence lower sea level. With no polar landmass, sea level will rise. Ocean–continent distribution also has profound effects on the circulation pattern of the oceans and therefore on global climate. The tectonic reshuffle of continent and ocean was probably the biggest single factor in developing the icehouse Earth conditions of the past 3–5 million years.

(4) Climate change and the natural oscillation between glacial and interglacial conditions. This affects the amount of water locked up as land ice or released back to the oceans in liquid form.

It is only this fourth trigger that is capable of effecting relatively rapid changes in sea level, such as those that have closely followed climate change over the past 3 million years at least. The rapid sea-level rise we see today, however, is very largely due to human-induced global warming. This leads to excessive melting of water that is locked up in glaciers and ice sheets on land and its direct addition to the ocean reservoir.

Models and controls

Global warming and its anthropogenic drive are not in doubt. What is remarkable is that human activity alone has been able to affect, at least in some marginal way, such a huge and complex ocean–climate system. Atmospheric levels of carbon dioxide have increased by 31 per cent since 1750, following widespread industrialization and an exponential increase in our burning of fossil fuels. Such large changes are relatively easy to measure, but the task of computing small variations in global mean temperature, set against the noise of large seasonal and decadal cycles, is really quite challenging, especially for the oceans.

The immense complexity of the natural system and its ocean regulator makes it enormously difficult to predict the Earth's continued response to increased greenhouse gas emissions. Only very recently has it become possible to make sophisticated computer models that realistically couple the atmosphere and ocean. These models show the pattern of atmospheric winds, the track of ocean currents, and the resulting transport of heat and gases around the world. They also show the extent of sea ice and land vegetation cover. Very slowly, we are beginning to make models that can *help* us predict future climate change.

Even if we have international consensus on the nature of global warming, we still have much to learn before we know how to respond. Reduction in greenhouse gas emissions, especially by the industrialized world, seems to be a logical first and immediate step. Methods for pumping carbon dioxide back into subsurface reservoirs have been piloted with success. An interesting new method of enhancing the natural transfer of carbon dioxide to the oceans is to stimulate phytoplankton growth by seeding the sea surface with dissolved iron. However, for some nations a warmer climate would bring very obvious benefits for agricultural production, energy consumption, and so on. Global warming is certainly one of the largest environmental issues facing humankind. It is not surprising, therefore, that there is still far from unanimous agreement about just how we should tackle it.

Chapter 7
Evolution and extinction

Life is inextricably bound with the oceans, from its first origins to its blossoming into the rich variety we know today. The passage of life through the different eras of ocean history is marked by evolutionary divergence and episodes of mass extinction, when up to 80 per cent of the planet's species were wiped out. Evolution has been painstakingly slow—from single-celled to multi-celled organisms, from asexual replication to sexual reproduction with associated mutations, and from soft-bodied creatures for which we have sparse fossil evidence to protective hard parts that are a thousand times easier to preserve. Most of this evolution and extinction has been played out in the oceans, to which life was confined for the first 3.6 billion years (or 90 per cent) of its existence. The changing nature of Earth's atmosphere and climate, fluctuations in the chemistry of seawater, the rise and fall of sea level, and the drifting of continents have all had different and profound effects. This chapter explores the likely origins of life, its evolution and marked changes, and the major extinction events that have punctuated this progress. The principal focus is the role of the oceans in the history of life.

Origins and early evolution

Although there is no single accepted model for the origin of life, there is general consensus that life began in the oceans and that water is a primary condition for its existence. This is why the presence of water on extraterrestrial planets is considered so vital in our search for life on other worlds. Not long after the first oceans formed and conditions stabilized, the first life appeared, probably sometime between 4 and 3.8 billion years ago. Fossil evidence for the first appearance of life, based on primitive microbial remains in marine sedimentary rocks from both Western Australia and South Africa, yields dates from 3.6 to 3.3 billion years ago.

Wherever in the ocean life began, those first single-celled organisms—primitive archaea and bacteria—faced a harsh environment of excess heat, damaging ultraviolet radiation, and bombardment from space. The early ocean waters were anoxic and acidic, with no free oxygen but rich in carbon dioxide, hydrogen sulphide, and iron. Different bacteria evolved to take advantage of alternative sources of hydrogen—some utilized methane in the early atmosphere, some the hydrogen sulphide from submarine volcanic emanations. Before long, new forms arose that were able to release hydrogen from water. These were the more complex cyanobacteria or blue-greens. They contained the complex molecule chlorophyll, which they used to harness the Sun's energy directly in the process of photosynthesis. By dissociating hydrogen from the water molecule they released oxygen into the ocean and atmosphere.

This was a clear evolutionary advantage for the blue-greens because of the evident abundance of both water and sunlight. It was also a significant development that was to have a profound effect on all life that followed. From about 2.45 billion years ago the levels of free oxygen in the atmosphere began to rise, albeit very slowly. Some of the oxygen is present as the ozone molecule, which provides a protective screen cutting out harmful ultraviolet rays from reaching the Earth's surface.

The first 2 billion years of life in the oceans was as nothing more than single-celled organisms, or *prokaryotes*, having no clear differentiation of parts within the cell and reproducing by simply splitting in two. Perhaps ocean conditions were relatively stable and unfavourable to change during this period. However, from around 2.1 to 1.8 billion years ago, there were two evolutionary breakthroughs that transformed life on Earth. The first was the development of cells having a nucleus—a core of material containing all the DNA instructions for life and reproduction (chromosomes or gene sets). These organisms, known as *eukaryotes*, gave rise to all other forms of life. The second was the advent of sex, which allowed the mixing of gene sets from two parents, via egg and sperm cells, rather than simple transfer from one. This greatly increased the potential for genetic variation by mutations, and hence accelerated the rate of evolution.

There was a limit, however, to evolutionary variety and size with only a single cell. It may have been the harsh ocean chemistry that constrained change, but that was slowly altering. Oxygen was building up near the surface, while the depths remained sulphidic, and there was an increase in bio-essential elements (Fe, Mo, Cu, Zn, and Cd). Two new evolutionary directions were tried out around 1.4 billion years ago. One was to form a colony of constituent cells with a coordinated life plan—the sponges of today are representatives of this very ancient line. The second and more successful development was to unite into true multicellular organisms. For the first time there was the potential to differentiate groups of cells according to function—a nervous system, sensory cells, reproductive parts, muscle fibres. The stage was set for great experimentation in life form.

Garden of Ediacara

In the Ediacaran Hills, part of the Flinders Ranges of South Australia, there are remarkable fossil impressions of the first known community of multicellular organisms, completely unlike anything that has lived before or since. They include large fern-like fronds, indentations of creatures that resembled jellyfish, soft corals, segmented worms, and ancient arthropods—all completely soft-bodied. Ripple marks in the sandstone and evidence of microbial mats suggest a shallow marine environment, presumably with a water chemistry and temperature that favoured such experimentation.

A new geological period, the Ediacaran (630–541 million years ago), was created for this time of fundamental change in ocean life. Fossils of the same period are now known from Russia, China, Namibia, Canada, and the UK—multicellular life had arrived. It seems likely that the oceans and atmosphere had begun to assume a chemistry similar to that of today. Atmospheric oxygen increased dramatically during the Ediacaran, reaching 15–20 per cent; the oceans, too, became fully oxygenated throughout the water column for the first time.

Ocean life explodes

Two events mark the onset of the Phanerozoic oceans 541 million years ago. The first is known as the 'Cambrian explosion' and heralded the appearance of a diversified and relatively advanced flora and fauna, which rapidly colonized the Cambrian seas. These included calcified algae, foraminifera and sponges, hard-shelled bivalves, gastropods and echinoderms, corals and sea lilies, together with now extinct but highly successful groups such as trilobites and graptolites. The second was the sudden evolution of hard parts—variously made of lime (calcium carbonate), silica, or chiton (a hard organic polysaccharide). Such mineralized skeletons are 1000 times more easily preserved as fossils than is soft tissue. Ocean chemistry had evolved still further—oceans were both fully oxygenated and less acidic, thus allowing for the preservation of hard skeletal material.

From the Paleozoic Era onwards, therefore, we have a much clearer image of how evolution proceeded (Figure 23). It was during the early Paleozoic that a quiet but fundamental experiment took place—the evolution of a backbone. The creature that first

Geological period	Date (million years ago)	Characteristic life
QUATERNARY	0 – 2.6	Mammoths, sabre-toothed cats, modern humans
NEOGENE	2.6 – 23	Grasses; hoofed mammals; rodents; snakes; first hominids
PALEOGENE	23 – 66	First whales; coral reefs; early ungulates; primates
CRETACEOUS	66 – 145	Calcareous plankton; flowering plants; placental mammals
JURASSIC	145 – 201	Modern fishes; early mammals; first birds
TRIASSIC	201 – 252	Scleratinian (stony) corals; reptiles; first dinosaurs
PERMEAN	252 – 298	Sponge-bryozoan reefs; advanced mammal-like reptiles
PENNSYLVANIAN (late Carboniferous)	298 – 323	Conifers; winged insects; first mammal-like reptile
MISSISSIPPIAN (early Carboniferous)	323 – 359	Seed-bearing plants; giant land-scorpions; first reptile
DEVONIAN	359 – 419	Large land plants; ammonoids; lungfish; sharks; amphibians
SILURIAN	419 – 444	Vascular plants; jawed fish; early land animals
ORDOVICIAN	444 – 485	Tabulate corals; jawless fishes; spores of land plants
CAMBRIAN	485 – 541	Arthropods; Burgess Shale fauna; first chordate; explosion of life forms
NEO-PROTEROZOIC	541 – 1,000	Ediacara fauna
MESO-PROEROZOIC	1,000 – 1,600	Multicellular organisms; Ediacara fauna
PALEO-PROTEROZOIC	1,600 – 2,500	Sexual reproduction, Eukaryotes evolve
ARCHEAN	2,500 – 4,000	Archaea; bacteria; cyano-bacteria; stromatolites, Abiogenic origin of life
HADEAN	4,000 – 4,600	

23. Evolution and events timescale.

displayed evidence of this structure was *Pikaia,* a flattened worm-like fossil found in the Burgess Shale of the Canadian Rockies. Not long after that the first proto-fishes appeared in the Silurian seas some 425 million years ago. Once the true fishes had evolved and mastered the art of swimming, they were set to conquer the seas.

It was at about this time, too, that the singular dominance of the aqueous realm as home to all life on Earth began to be challenged—after more than 4 billion years. First bacteria, algae, lichens, and fungi moved from the shoreline fringe to colonize the continental interior, and then the higher plants. Before long the continents were as green as the seas were blue. The advent of this colonization by land plants was to have a major and irreversible effect on the whole global environment. Pioneers of the animal world soon took to the land themselves.

Paleozoic mass extinction

Evolution and extinction have always been closely intertwined. At certain times and places in the past, conditions have favoured adaptation and change, great radiations of new creatures such as occurred in the Cambrian explosion of life, and at the beginning of each new geological era. Other times have witnessed the rapid and simultaneous demise of large numbers of species, genera, and whole families. These are the mass extinction events, including the ones that brought both Paleozoic and Mesozoic Eras to an end (Figure 24).

The largest of all took place around 250 million years ago at the end of the Permian period. An estimated 50 per cent of all known families disappeared forever, which equates to around 96 per cent of all species on Earth—most of which were still ocean-bound at that time. Before this, marine life had been a rich cornucopia of species quite unlike those of today but equally varied, forming complex food webs that had been evolving for over 100 million years. Almost all

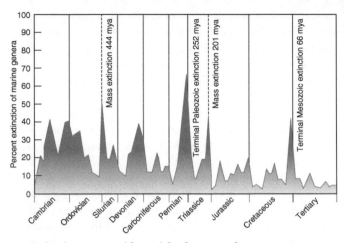

24. Extinction events, with special reference to the oceans.

solitary corals, sea lilies, and encrusting bryozoans died out, together
with nearly all the 160 known species of brachiopods. The last
of the trilobites crawled over the seabed, and the last graptolites
disappeared from the surface waters. This was very largely an ocean
extinction, but on land too, the great coniferous forests of Laurasia
and Gondwanaland were wiped out, taking with them a host of
insects and other small creatures for which we have little record.
Only very few of the larger mammal-like reptiles survived.

This great mass extinction event was quite unprecedented in
scale and effect. It was drawn out over 8 million years of rapid
decline and a shorter 80,000-year period of catastrophic species
demise. We can infer that two complex environmental factors
were responsible. The first was the slow fusing together of
continents during the Permian period to form a single great
landmass known as Pangaea. This had several consequences:
(a) sea levels fell to an all-time low reducing habitat diversity and
promoting ecological instability; (b) the escape of gas hydrates
and oxidation of carbonaceous deposits, also a result of lowered
sea level, released large quantities of carbon dioxide into the

atmosphere thus causing global warming—atmospheric oxygen levels plummeted; (c) continental climates were extreme, severely stressing plant and animal life on land.

The second was an intense period of volcanic activity, for which we have good evidence in the Siberian Traps, which represent enormous outpourings of lava. Immense clouds of volcanic ash may have temporarily blotted out sunlight and led to a 'nuclear winter'. But ultimately the result of vast emissions of carbon dioxide and other gases led to global warming and, perhaps more importantly, to poisoning of the atmosphere and oceans with fluorine, acid rain, and trace metals. Low atmospheric oxygen would also transmit to the oceans.

Pushed to their limits by such a combination of environmental factors, the natural decline of species increased many times over. The loss of certain plants affected a host of animals that fed on them and, in turn, the predators further up the food chain. Competition inevitably increased as favourable habitats diminished. Diseases may have spread more aggressively between animal groups as once-isolated landmasses came together. In the final analysis, it was an unparalleled biological disaster from which the world took many million years to recover.

Tethys Ocean revival

The Mesozoic Era heralded a completely different world. The single supercontinent that had formed in the Permian period proved unstable and began to break apart. A series of east–west rifts opened up a narrow equatorial ocean right across the heart of Pangaea. This was the Tethys Ocean that grew and grew and came to dominate the world for the next 250 million years. Sea levels rose to the highest they had ever been, global temperature rose, more rifts appeared in the continents, and arms of this warm new Tethys Ocean fingered into the continental interior.

Such a benign ocean world was an ideal nursery for the revival of life. New types of microscopic plant life took hold of the phytoplankton world. Coccolithophores, diatoms, and dinoflagellates all came to prominence. There was a parallel explosive radiation of zooplankton to feed upon such bounty—lime-secreting foraminifers and siliceous diatoms at the forefront. Marine invertebrates capitalized on the ocean rejuvenation that was now underway with an exuberant radiation in form. Molluscs proliferated—from the oysters and clams that hugged, burrowed, or attached themselves to the substrate, and great reef-like mounds constructed almost entirely by rudist bivalves, to the highly effective free-swimming shellfish (coiled ammonites and bullet-shaped belemnites). At one period a vast deep-water sponge reef, about one-and-a-half times as long as the Great Barrier Reef today, stretched along the northern Tethys shelf seas. Later still, new types of corals and bryozoans appeared in warm shallow seas all around the ocean.

While dinosaurs proliferated on land, so marine reptiles became the top predators at sea: saltwater crocodiles, paddle-finned long-necked plesiosaurs, and big-headed short-necked pliosaurs, some growing up to 17 metres in length, about twice as long as a bus, with powerful jaws and gigantic teeth. Perhaps most highly evolved were the fast-swimming, dolphin-like ichthyosaurs.

Almost unnoticed amid the weird and wonderful profusion of the time were several key developments, on land rather than at sea, which were to form the cornerstone of future life. First was the evolution of mammals from one of the therapsid group of advanced mammal-like reptiles in the late Triassic Period. Mammals remained small and lay low beneath the overwhelming dominance of reptiles. Second was the evolution of birds, most probably from small theropod dinosaurs in the early Jurassic, like the much-heralded *Archaeopteryx*. And third was the appearance of angiosperms (flowering plants) on land in the early Cretaceous.

Mesozoic mass extinction

The fecundity of the Mesozoic oceans was brought to an end by the mass extinction event at the end of the Cretaceous period, some 66 million years ago. The planktonic world underwent severe decline and dramatic change, with knock-on effects upwards through the food chains. Ammonites and belemnites disappeared forever. With them went roughly a quarter of all species of crocodiles, turtles, and fish, and almost all the top-predator marine reptiles. On land, the long decline of the dinosaurs was completed. Altogether about 20 per cent of known families and an estimated 50 per cent of all species became extinct. The duration of this event is known to have been relatively short, but not instantaneous.

Literally hundreds of theories have been proposed for the death of the dinosaurs—some with an element of sense, many more quite fanciful. Much advocated at present is the idea that a giant asteroid, measuring 10 kilometres in diameter and weighing in excess of 4 million tonnes, struck the Earth somewhere in the Gulf of Mexico region. The ghost of a giant ring-like structure has been seen on remote images of the area and called the Chicxulub crater. Concentrations of iridium, an element common in some meteorites, have been found across the Cretaceous–Tertiary (KT) boundary at different places around the world, together with a highly stressed type of quartz grain. But, in reality, the evidence is meagre. Such catastrophist theories have difficulty explaining the fact that many plants and animals were barely affected—most land plants, for example, survived. Most molluscs, sharks, bony fishes, placental mammals, and all amphibians were also completely unscathed.

There is abundant evidence, however, for a combination of environmental drivers. As with the end Permian extinction, there were large outpourings of lava just preceding the KT boundary event—evidenced by 500,000 square kilometres of Deccan Traps in India. Extremely high sea levels and a warm climate had endured through 30 million years of the late Cretaceous. Then, sea levels began to fall and temperatures dropped dramatically. Coastal habitats diminished and land bridges opened up—the spread of diseases between animal groups, already subject to enhanced competition, would have increased. There is further evidence of long-term decline in many groups of dinosaur, and in the extinction of certain marine fauna several millions of years prior to the boundary. Detailed microfossil records across the KT boundary in marine records drilled beneath the oceans reveal a rapid but stepped species extinction.

It is difficult to avoid concluding, therefore, that the KT extinction event was due to a complex combination of environmental factors, including marked changes in ocean chemistry, which triggered extreme biological stress.

Cenozoic ocean world

Once again the oceans helped nurse the world back to vitality. Continental drift continued to change the distribution of land and sea. The Tethys Ocean was slowly closing—and finally disappeared for good around 5.5 million years ago. The Atlantic and Indian Oceans were expanding, the Antarctic continent drifted across the South Pole, and the world began to look much as it does today. New plankton evolved together with a plethora of colourful fishes and modern corals. Oysters, mussels, barnacles, cockles, and razor clams adorned the rocky and sandy shorelines, while an altogether different underworld colonized the deep oceans. This included unusual chemosynthetic ecosystems around black-smoker vents on mid-ocean ridges, cold-water coral communities that thrive without direct sunlight, and a host of remarkable bioluminescent organisms capable of generating their own light source.

One significant evolutionary development of the Cenozoic Era was the return to the sea of a mammalian line that had become entirely land-based. This was represented by hyena-like creatures, known as *Pachyaena*, which scavenged along the shoreline 45–50 million years ago. Fossil remains show progressive and rapid evolution from these strand hunters to wading 'whales', to paddle-swimmers, and thence into tail-swimming, streamlined, fully marine mammals of the cetacean family. The first to appear were the toothed whales (dolphins and porpoises), whereas the baleen whales evolved some 10–15 million years later.

Chapter 8
Marine web of life

The oceans now cover 71 per cent of the world and over 90 per cent of its habitable living space. They provide vital nutrients, dissolved gases, and mineral salts. They are rich in oxygen and their surface is bathed in an endless supply of sunlight. Water temperatures are far more constant than those on the continents and they are responsible for over half the world's primary productivity.

Life exists through the cycling of matter and specific chemicals, the consumption of food, and the exchange of energy. Every marine organism is affected by its surroundings. Temperature, salinity, water pressure, and the availability of sunlight and of nutrient chemicals are some of the physical factors that influence ocean life. The assemblage of organisms that inhabit any one ecological niche also has a profound influence on every member of the group. There is competition for living space and for food, just as there is an urge to reproduce. But the endless quest for food is the foremost need of all living creatures. Different species play their own unique role in the cycles, complex webs, and food chains of the oceans. Each of the six kingdoms into which we classify life is well represented in the marine realm (Table 6). This chapter provides a small introduction to the much larger topic of marine biology.

Table 6 The six kingdoms of life as represented in the oceans

Archaea	Prokaryotes, single-celled organisms, one of the oldest, most primitive, and most abundant life forms, especially adapted to harsh environments
Bacteria	Prokaryotes, single-celled organisms, one of the oldest, most primitive, and most abundant life forms, more or less ubiquitous
Protista	Eukaryotes, single-celled organisms with well-defined differentiation of nucleus and organelles, catch-all group for animal-like, plant-like, and intermediate forms
Fungi	Eukaryotes, mostly multicellular and lacking independent movement, heterotrophic decomposers of dead and living organic material
Plantac	Eukaryotes, multicellular, and lacking independent movement, autotrophic (primary producers) with an ability to produce their own food through photosynthesis
Animalia	Eukaryotes, multicellular, and mostly showing some degree of independent movement, heterotrophic (consumers)—feeding on other organisms

Biozones and habitats

Ocean space is generally subdivided into *biozones* according to its particular physical attributes (Figure 25). In the first instance, a clear distinction can be made between the *pelagic* province, which includes the whole of the water column and its host of nektonic and planktonic organisms, and the *benthic* province or ocean floor, scoured and inhabited by benthic organisms. A fundamental division is also made between the *neritic* zone of shallow water that overlies the continental shelves, and the *oceanic* zone of the open ocean beyond the shelf break. This can be further subdivided on the basis of depth into littoral (or intertidal), sub-littoral (continental shelf), bathyal (continental slope), abyssal (deep-ocean

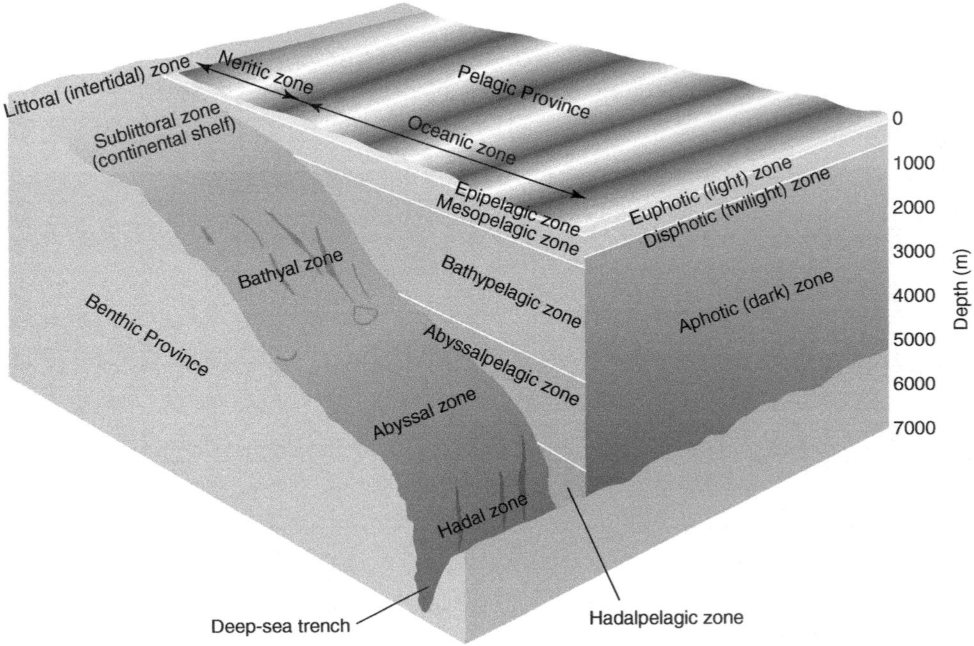

25. **Marine biozones and habitats.**

Oceans

Depth (m)

0
1000
2000
3000
4000
5000
6000
7000

Euphotic (light) zone
Disphotic (twilight) zone
Aphotic (dark) zone

Pelagic Province

Oceanic zone

Epipelagic zone
Mesopelagic zone

Bathypelagic zone

Abyssalpelagic zone

Hadalpelagic zone

Neritic zone

Littoral (intertidal) zone

Sublittoral zone
(continental shelf)

Bathyal zone

Abyssal zone

Hadal zone

Benthic Province

Deep-sea trench

basin), and hadal (deep-sea trench) zones. Depth is one of the key controls on the availability of light in the ocean and hence on the nature and distribution of biota. Deep-sea habitats and ecosystems are one of the least known on Earth.

The broad climatic belts expressed on land, such as Arctic tundra, temperate grasslands, and tropical rainforest, have their muted but equally significant expression at sea. Floating ice and winter darkness make the *polar* seas one of the harshest climates for any life to survive, thus forcing some of the most remarkable seasonal migrations on the planet. The *temperate* zones are mild in climate but still markedly seasonal. This drives a strong ocean circulation pattern, overturn of water masses, and recycling of nutrients. The result is a spring and summer productive bonanza, which quite literally feeds the world—an explosive series of planktonic blooms that support many different colonies and ecosystems. An equatorial *tropical* zone of permanently warm seas girdles the Earth. Although its productivity is distinctly patchy and less than that of the temperate seas, it sports some of the most diverse and exotic of all marine habitats, including coral reefs and mangrove swamps.

The possible choice of a home or habitat in the ocean seems almost endless. At first sight it appears strange that animals might choose the icy polar waters in preference to balmy tropics, or swap the bounty of temperate shelves for a near starvation diet on the deep-sea floor. But that is the marvel of competition and adaptation by which all organisms have evolved over countless aeons. Indeed, the nutrient-rich temperate seas support very large numbers of organisms and a whole series of complex food webs. It may be easy to eat and find a mate but it is equally easy to become somebody else's lunch. Coral reefs may provide a smorgasbord of tasty entrées and all manner of places to hide, but also an equal number of predators with that special adaptation to penetrate even the most secure of hiding places or most impenetrable of armour. Set against this backdrop of fierce competition, it becomes clear why colonization of new and harsher habitats, accompanied

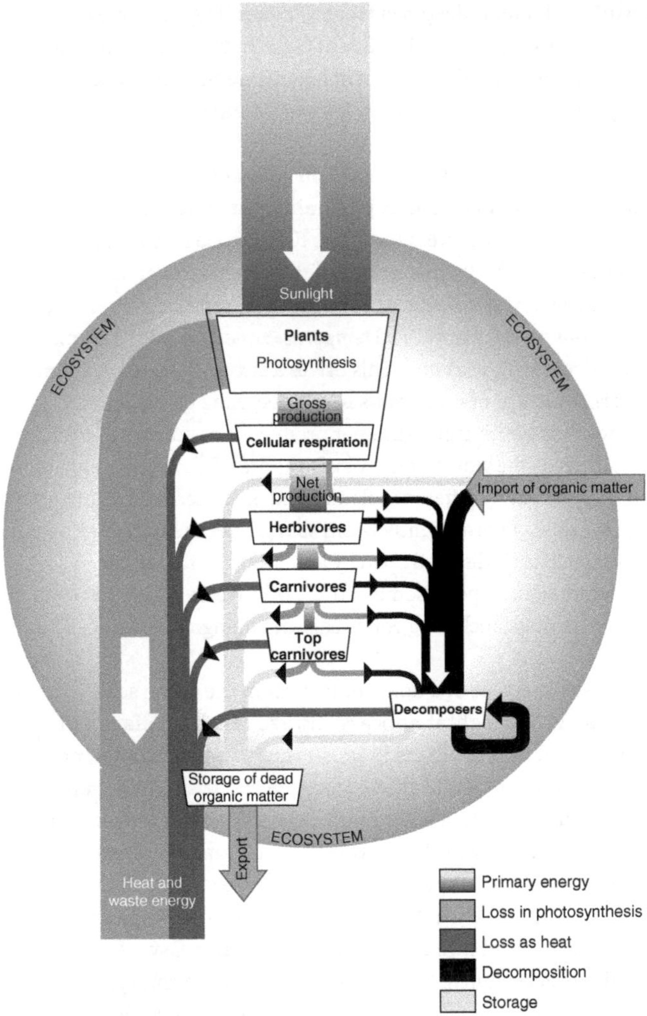

Sunlight

Plants
Photosynthesis

Gross production

Cellular respiration

Net production

Herbivores

Carnivores

Top carnivores

Import of organic matter

Decomposers

Storage of dead organic matter

Export

Heat and waste energy

ECOSYSTEM

Oceans

	Primary energy
	Loss in photosynthesis
	Loss as heat
	Decomposition
	Storage

26. Ocean life cycles and webs.

by appropriate adaptation to survive, becomes an attractive alternative to the 'urban' rat race.

Whatever home is eventually chosen—rich suburb, poor shanty town, or remote farmhouse—each organism must find its own particular niche within that habitat in order to survive (Figure 26). In an ecological context, an organism's niche is its occupation. For example, barnacles stick to wave-pounded rocks that few other creatures could survive, filtering food from the surf, while sea cucumbers crawl quietly across the muddy seafloor hidden from light, scavenging and recycling the organic waste from others. The precise niche is determined by a wide range of abiotic and biotic factors to which the organism responds. The more unique the niche and the less it overlaps with those of other organisms, the more successful an organism will be.

The world of plankton

The soup kitchen of the ocean world is the plankton—a multitude of microscopic and tiny organisms that live near the sea surface in the pelagic province. Each year over 6 million million tonnes of *phytoplankton* grow wherever light will penetrate, harnessing solar energy and chemical nutrients in seawater (Table 7, Figure 27).

There are at least 6000 different species of diatom. These miniature organisms create a protective home in silica glass boxes of all shapes and designs. Most store food reserves as droplets of oils and fatty acids, which also help maintain buoyancy and hence a prime position near the sea surface. Diatoms are the dominant phytoplankton in cold and temperate waters, as well as along many shores and coastal inlets. Another group, the silicoflagellates, are most abundant in cold polar waters.

In tropical seas and the open ocean coccolithophores abound. They make their complex spherical homes of tiny patterned discs (coccoliths) from calcium carbonate. They have two thread-like

Table 7 Primary production statistics

Quantity: 500–750 gC/m²/year
Ocean: shallow inlets, kelp beds, coral reefs
Land: tropical rainforests, freshwater swamps

Quantity: 250–500 gC/m²/year
Ocean: coastal upwelling systems, deep estuaries
Land: intensive agricultural areas, lakes

Quantity: 50–250 gC/m²/year
Ocean: continental shelves and some marginal seas
Land: temperate forests, grasslands, croplands

Quantity: < 50 gC/m²/year
Ocean: open ocean areas
Land: deserts, steppes

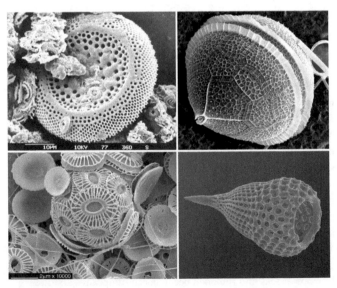

27. **Plankton community.**

flagella, which they use as a weak form of locomotion, primarily maintaining their position in the sunlight. Dinoflagellates are other members of the tropical phytoplankton, some of which glow with a blue-green bioluminescence at night. Some species have opted for a symbiotic life within the bodies of jellyfish, corals, and molluscs. Their task is to supply food, while their host provides protection, carbon dioxide, and other nutrients. Other dinoflagellates have taken up a parasitic existence, living in the intestines of marine crustaceans.

Formerly unnoticed in the phytoplankton, because of their ultra-microscopic size, are the cyanobacteria. More than half a million of their individual cells would fit on to a single pinhead. Yet we now know that they can account for 80 per cent of primary production at the ocean surface. In shallow waters, where sunlight easily penetrates, they are also found carpeting the seafloor as rather slippery, dense green mats. Those forms that secrete calcium carbonate are architects of cabbage-like stromatolites—living representatives of an ancient life form.

Everywhere that primary producers live, they are accompanied by an equally diverse *zooplankton* community of tiny herbivores and carnivores. Protozoans are all single-celled organisms that commonly build tiny sculptured shells, about the size of sand grains. They include the multi-chambered shells of foraminiferans, made from calcium carbonate, spiky glass spheres of radiolarians, and helmet-shaped tintinnids. These organisms extend fine radiating strands of their own cell material through tiny pores, which they use for capturing prey as well as for locomotion.

The chief consumers of both phytoplankton and other zooplankton are copepods. They grow a little bit larger, up to 2 millimetres in length, hunt more ferociously, and consume more rapidly than the smaller protozoans. They are tiny elongate shrimps with broad waving antennae, which have developed highly efficient means of grazing. They beat their legs furiously, setting up a steady flow of water that directs the floating plankton towards feeding cups spaced out along their limbs.

In the Southern Ocean, high plankton productivity is related to upwelling of nutrient-rich waters along the polar front that separates different water masses. The diatom blooms at these high latitudes generally begin in the early summer months, when pack ice begins to melt and the hours of sunlight increase. As the phytoplankton bloom they attract krill—open-water shrimps that grow to several centimetres in length and multiply rapidly. Single swarms can be hundreds of square metres in area, 5 metres thick, and have been estimated at around 2 million tonnes in weight. Most of the eighty-six known species of krill are bioluminescent, so that these incredibly dense swarms produce an eerie glow at night.

The marine environment

Light availability, temperature variation, saltwater chemistry, pressure, and density are the principal physical parameters of the marine environment. These are the factors that all marine organisms must learn to live with if they are to survive.

Sunlight is vital for all photosynthesizing organisms, which have therefore developed a multitude of adaptations for maintaining their position in the photic zone—less than a few metres from the surface in cloudy coastal waters or down to 200 metres in the clear open ocean. The majority of heterotrophic organisms must similarly adapt to live where food is plentiful. There is also a diverse range of creatures which have adapted to life without sunlight, below even the twilight zone.

Most of the solar energy absorbed by seawater is converted directly to heat, and water temperature is vital for the distribution and activity of life in the oceans. Whereas mean temperature ranges from 0 to 40 degrees Celsius, 90 per cent of the oceans are permanently below 5°C. Most marine animals are ectotherms (cold-blooded), which means that they obtain their body heat from their surroundings. They generally have narrow tolerance limits and are restricted to particular latitudinal belts or water depths. Marine mammals and birds are endotherms (warm-blooded), which means that their metabolism generates heat internally thereby allowing the organism to maintain constant body temperature. They can tolerate a much wider range of external conditions.

Coping with the extreme (hydrostatic) pressure exerted at depth within the ocean is a challenge. For every 30 metres of water, the pressure increases by 3 atmospheres—roughly equivalent to the weight of an elephant. At 1 kilometre depth, the pressure is 100 atmospheres—or the equivalent of thirty-three elephants piled up on just one square metre. For most surface-dwelling species, these crushing pressures would be untenable, but there is a rich and diverse community of organisms that populates the abyssal seafloor. This is because it is gases not liquids that are highly compressible and all marine creatures are composed mainly of water and contain no gases. They are apparently quite unaware of the oppressive hydrostatic pressures under which they live. Rather more specialized adaptations are required, however, for those fishes and whales that can live at the surface and then dive into the deep sea in search of food.

Movement underwater

Several different styles of movement are used by marine organisms. These include floating, swimming, jet propulsion, creeping, crawling, and burrowing. A few species have also learned to hitch a ride on others, or to catch the wind and sail across the sea surface. Yet other animals choose a sedentary existence, at least for most of their life. The particular physical properties of water that most affect movement are density, viscosity, and buoyancy. Seawater is about 800 times denser than air and nearly 100 times more viscous. Consequently there is much more resistance to movement than on land, as anyone will know who has ever tried to run through even waist-deep water.

Floating and drifting is the preferred option for many zooplankton that reside amid phytoplankton pastures. Others display a much greater range of movement. Some planktonic animals only 1–2 millimetres in length actually travel several hundred metres each day, swimming up to the surface to feed at might. At dawn they sink back down in an effort to avoid predation. The double journey would be like a person swimming 700 kilometres! Even jellyfish use a weak form of jet propulsion, squeezing water from their floating sacks to effect motion.

Most large marine animals, including all fishes and mammals, have adopted some form of active swimming (Figure 28). Swimming efficiency in fishes has been achieved by minimizing the three types of drag resistance created by friction, turbulence, and body form. To reduce surface friction, the body must be smooth and rounded like a sphere. The scales of most fish are also covered with slime as further lubrication. To reduce form drag, the cross-sectional area of the body should be minimal—a pencil shape is ideal. To reduce the turbulent drag as water flows around the moving body, a rounded front end and tapered rear is required. The resultant shape, taking into account all three types of drag, is the torpedo form of a tuna fish—this is the fastest swimming of all fishes.

Fins play a versatile role in the movement of a fish. There are several types including dorsal fins along the back, caudal or tail fins, and anal fins on the belly just behind the anus. Operating together, the beating fins provide stability and steering, forwards and reverse propulsion, and braking. They also help determine whether the motion is up or down, forwards or backwards.

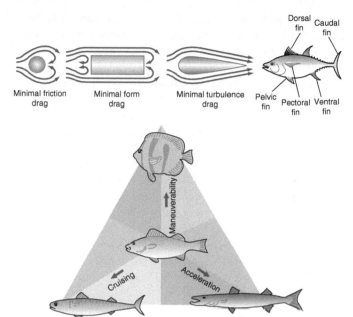

Minimal friction
drag

Minimal form
drag

Minimal turbulence
drag

Dorsal
fin

Caudal
fin

Pelvic
fin

Pectoral
fin

Ventral
fin

Maneuverability

Cruising

Acceleration

28. Movement in fishes.

Several fishes have learned to glide through the air. For the fast-swimming flying fish it is probably the most important evolutionary adaptation that has enabled it to survive through millions of years of predation by equally swift hunters. Launching into the air requires the fish to swim upwards at high speed, then as it begins to leave the surface it thrusts strongly from side to side with its enlarged caudal fin, and extends its broad wing-like pectoral fins for complete lift off. These fishes can leap as high as 10 metres, soaring through the air at speeds of 60 kilometres per hour and travelling 250 metres in a single flight. The fast whirring action of their caudal fins as they return to the sea allows them to take off again and make several long flights in a row, even turning a near right angle in mid-flight.

For every creature that swims or floats, there are many more that have opted for a slower or more sedentary existence on or around the seafloor. These include bacteria and fungi that clean and recycle waste, photosynthesizing algae within the photic zone, as well as a host of larger animals. Those that burrow into the sediment for protection (worms and some shellfish) as well as sessile anemones, corals, sponges, and bryozoans, simply play a waiting game for food to fall to the seafloor or wash up along the shoreline.

Other bivalves, including oysters, clams, scallops, and mussels, as well as a variety of sea snails (gastropods) and sea slugs, creep very slowly over the seafloor. The sea slug uses a similar method of movement as it munches its way across a diet of seaweed, or an animal host to which it has become attached. Crustacean scavengers—lobsters, crabs, crayfish, and shrimps—use four of their five pairs of jointed legs to walk about on the ocean floor, a clumsy almost comical movement like heavily armoured knights.

Heightened senses

Marine animals have developed a heightened awareness of several different sources of sensory information. Sight and sound are important for some, whereas for others it may be pressure differences, smell, touch, taste, or even gravity or magnetic and electrical fields.

There are a wide range of adaptations for vision under the sea. The flatworm has an organ analogous to an eye with tiny pigment spots, which allow it to distinguish light and dark. As there are two receptors, it can also determine the direction of the light. The giant Pacific scallop has a series of receptors rimming the fleshy periphery of its body that give a simple sense of light and dark, perhaps warning the shellfish to clamp tight shut when the shadow of a predator approaches. Most fishes have eyes set on opposite sides of the head and use them independently for monocular vision. By weaving their head and body from side to side as they swim, each eye can scan a wide range. Some fish can swivel their eyes forwards and so achieve limited binocular vision. Having no muscles to alter the shape of the lenses, they must move the entire lens backwards to see distant objects and forwards to view near objects, in exactly the same way as we focus a camera.

Chemical stimuli are easily transmitted through water and so form an important source of information. Sea anemones begin to reach out their waving tentacles and grasp at any suitable food particles that come within their chemical range. A great many other creatures have well-honed chemoreceptor cells—from primitive protozoans and colonial sponges, to salps and sea cucumbers, carnivorous molluscs and crustaceans, and all types of fish.

Most fishes have a well-developed sense of smell, some much more so than others. The moray eel is a common reef resident, hiding in crevices by day and hunting by night, locating and tracking its prey largely by scent. The Nassau grouper has a double nasal opening giving it a highly efficient olfactory sense. It is well known that sharks have an uncanny sense for the least trace of blood within a wide circle of their location, as well as excellent vision up to about 15 metres. They can detect food or blood in the water at levels of only a few parts per hundred million.

All elasmobranchs—skates, rays, sawfish, and sharks—are able to sense electricity, through special organs (ampullae of Lorenzini) located over the top and sides of the head region. They are small sacks of sensitive tissue filled with a jelly-like substance that are connected to the surface via narrow ducts and pores. These ampullae can detect as little as one one-millionth of a volt of electricity. Since all animals involuntarily generate minute electrical charges as they live and move, the ability to detect and interpret such charges is a formidable weapon.

Such noisy silence

Far from being a silent, tranquil underworld, the sea is filled with a cacophony of sound. Animals of all types contribute clicks, wheezes, whines, clucks, rumbles, scrapes, and bumps. All these sounds are transmitted by compressive wave motion far more effectively underwater than in air. As a result sound travels twice as fast in water and about four times as far.

Fish, shellfish, and crustaceans have no vocal cords to help them emit sound, but use different adaptations. Fish of the grunt family grind together their upper and lower pharyngeal teeth emitting a sound not unlike that of feeding pigs. The croakers and drummers are also aptly named fishes for the sounds they produce by vibrating strong muscles attached to the swim bladder walls. The toadfish both grunts and whistles, the latter as long low-frequency bursts of sound used typically in the mating season. The pistol shrimp, though less than 5 centimetres long, can produce a very sudden and sharp retort by snapping together two parts of its outsized claw. This is highly effective for stunning small prey and also in defending their burrows. Fiddler crabs use their claws to rap out warning signals, and leg movements to produce a lower frequency honking sound that attracts female crabs in the vicinity. The variety is as intriguing as it is diverse.

The lateral line is a system of mucus-filled canals just beneath the surface of the skin that stretches along the length of many fishes and heightens their response to infrasonic vibrations. Fishes are therefore very aware of the least movement underwater, which is one reason that a whole school of many hundreds of fish is able to twist and turn in exact unison.

Sound is the principal sensory system used by larger marine mammals such as whales and dolphins. Dolphins emit high-frequency sounds, from audible clicks to ultrasonic vibrations of over 100,000 Hz, and then listen for the echoes. From these they construct an accurate image of their surrounds—the distance, direction, and movement of any objects, their size and shape, texture, and density, and even species of fish. The clicks are produced from the dolphin's large forehead region, by recycled compressed air, and the echoes are received over a broader area of the head including the inner ears. A large auditory lobe in the brain helps decode such a complex array of signals.

Feeding strategies

So many of an animal's adaptations for life in the sea are ultimately strategies for feeding and for avoiding predation. This is the harsh reality of survival, in which all animals tread the finest of lines between life and death. From the top predators to tiny zooplankton, each animal establishes its own ecological niche and then adopts a suitable feeding strategy. To achieve this necessitates the correct selection of tools and techniques, armoury and camouflage, and the decision to cooperate or to avoid conflict.

Eating low on the food chain is the most energy efficient means of nourishment for animal life. Size and speed of movement do not matter. From tiny zooplankton and slow-moving sea snails to the great blue whale, the principal determinant is locating enough to eat. Baleen whales are the ultimate in perfection and scale when it comes to harvesting a meal from the ocean plankton. The blue whale is the largest creature on the planet and eats a daily average of 4 tonnes of plankton. Mostly it eats krill, a small shrimp-like member of the zooplankton, but simply by opening its enormous mouth it swallows everything in its path. In place of teeth, there are large plates of baleen (a keratin substance) forming a tight mesh. When the whale's mouth is full of food and water, it closes and expels the water through the baleen plates straining out the krill and other plankton, which it then swallows. Humpback whales often feed by ascending slowly from below a rich patch of plankton, blowing a ring of bubbles. This acts as a bubble net, trapping the krill as a concentrated broth near the surface.

Other creatures prefer to settle down for most or all their life in just one place. This is an excellent way to conserve energy, not having to move but simply letting food and water waft towards them, perhaps helping it along by flailing cilia or waving tentacles. The chief problem is to find a suitable location where food will be naturally plentiful, and a suitable substrate. Polychaete worms and many bivalves burrow into soft muds and sand; other bivalves live on or bore into rock. Barnacles, the small hardy crustaceans whose conical encrusting shells are so familiar along every rocky seashore, attach themselves to almost anything—rocks, ship's hulls, and other marine animals including crabs, turtles, fishes, and whales.

Corals and other reef builders construct solid, three-dimensional edifices that may cover many square kilometres of seafloor. Individual coral polyps are tiny soft-bodied creatures that live in and over the little cups of calcium carbonate they have secreted. Typically, their tentacles are lined with nematocysts (stinging cells), and also have harpoon tips for firing poison. They also extrude long, fine threads to wrap around and ensnare prey, perhaps coated with a sticky substance, which further prevents escape. Some are shorter fleshy stalks covered in minute cilia that beat in unison, directing a flow of water and foodstuff over the polyp's mucus-lined interior. Equally effective is the polyp's common symbiotic relationship with photosynthesizing dinoflagellate algae, known as zooxanthellae.

Hunting and poisoning

For those animals that hunt for a living, locating and trapping a meal is a more or less constant pursuit, aided by a range of hunting equipment—teeth of all kinds and numbers, sharpened spikes, swords and tusks, electric shocks and loud stun guns, suckers and stings, and poisons that immobilize and kill.

The cruising fish and toothed mammals are some of the most voracious predators in the ocean and some of the most effective, capable of devouring large quantities of smaller prey in a matter of moments. There are some 250 species of shark, 35–40 of which are dangerous to humans. Their several rows of large, sharp, triangular teeth coupled with immensely powerful jaws, keen senses, and swift movement cast them among the top marine predators, alongside dolphins and killer whales.

For others, the hunting technique is more varied and creative. Many reef-feeding fishes have elongate snouts for prying into cracks and crevices. Sea stars use strong suckers to prise open shellfish, while the crown-of-thorns starfish has an anvil of small rasping teeth to grind and suck away at coral polyps. Sea spiders feed on the juices of cnidarians and other soft-bodied invertebrates that they extract with a long sucking proboscis. The bat ray swims slowly over the seafloor spurting water from its gills to stir up the sediment, thereby revealing any hiding animals. It then settles and sucks the helpless victims into its mouth. Electric rays and torpedo rays are among the 250 species of fish known to possess specialized organs capable of delivering painful electric shocks (from 8 to 220 volts), stunning or even killing their prey.

The art of poisoning is even more prevalent at sea than on land. Many species of five main invertebrate phyla and a large number of vertebrates regularly use venom, either as a defensive mechanism or for immobilizing prey prior to feeding. Several of the sedentary animal groups employ batteries of stinging cells (nematocysts) for both defence and attack, and often advertise the fact with bright colours—red moss sponge and fire sponge, strawberry anemone and hell's fire sea anemone, for example. Nearly all the 2700 varieties of hydroid also deploy rows of nematocysts along their waving tentacles.

Some of the most venomous creatures in the ocean include the box jellyfishes and sea wasps of northern Australia and the north Indian Ocean. These can kill a person within thirty seconds of being stung as a result of circulatory or respiratory failure. Among many different types of venomous fishes—dogfish, dragonfish, scorpionfish, stonefish, and pufferfish—the most dangerous of all are the superbly camouflaged stonefish, which can kill small prey in seconds and a human swimmer within a couple of hours, and the pufferfish whose toxin is 200,000 times more poisonous than the deadly plant extract used to tip poison arrows.

Sex at sea

Food and reproduction are the chief preoccupation of most living organisms, and the scale of reproduction in the ocean is quite phenomenal. An average litre of surface water from the ocean contains over half a million diatoms and other unicellular phytoplankton and many thousands of zooplankton. During the spring bloom, each cell divides and rapidly reproduces, asexually, easily yielding more than a billion progeny in a month.

Copepods are among the first 'larger' animals drawn to such spring blooms. As they graze and reproduce, sexually, so their numbers increase astronomically. After fertilization a female copepod can produce over a hundred eggs, and some species produce a new clutch of eggs every 4–5 days. New hatchlings take less than two weeks to reach maturity before they too can reproduce. Where food is plentiful, especially in the surface waters across broad continental shelves of upwelling areas, their concentration can reach 100,000 individuals per cubic metre of water. At such times, they probably rank among the most numerous animals on the planet. Each one can consume as many as 120,000 diatoms per day.

The ability to lay large numbers of eggs is a major advantage for species survival. Horseshoe crabs each lay 20,000 eggs and their forebears, who have changed very little in appearance through time, have probably been doing the same every spring for the past 500 million years. The average hake will lay 1 million eggs at a time, a haddock up to 3 million, and a cod as many as 9 million. The purple sea hare snail produces 20 million eggs, while an average oyster will lay 500 million eggs in a single year! The individual polyps of a large coral reef effect a multi-species synchronized mass spawning with the simultaneous release of billions of eggs and sperm. There is an amazing pinkish haze across the coral reef when this occurs.

There is enormous variety in how reproduction is achieved, and many species keep several options open. Corals, sponges, and sea anemones can reproduce by an asexual budding process, as well as by spawning. Ostracods can reproduce without the need of male sperm. A number of organisms have evolved a hermaphrodite lifestyle and can become either male or female as required by the sex ratio of the colony—limpets and giant clams, for example. Older giant clams (aged 6–8 years) become both male and female simultaneously, and so release both eggs and sperm.

To gain an advantage in sexual reproduction, many techniques are employed. Colour displays, bioluminescent signals, and the release of chemical pheromones are all common practice. Establishing and defending a territory for mating engages males of many species, from tiny worms, fishes, and cephalopods, to mighty tusked walruses. Building and protecting nests can involve both partners, while the jawfish lays her eggs in the male's mouth, where he incubates them until they hatch.

Ritualistic mating dances and courtships are often spectacular and intricate, and may be played out over several hours before mating or egg and sperm release occurs. Marine mammals, in particular, perform these elaborate routines. They are viviparous, in that eggs are fertilized and the embryo grows within the female body. Cetaceans (whales and dolphins) give birth to only one or two offspring who then travel with the larger pod for protection as they grow.

The oceans, however, remain a challenging environment for life, as over 99 per cent of progeny will never survive to maturity.

The deep-ocean habitat

Of all the marine communities, life in the deep sea is the most remarkable and still the least known. Light fades gradually through the upper ocean so that, even in the aphotic zone, below the level where photosynthesis occurs, there is still some vestige of light down to about 600 metres in the clearest of waters. This eerie twilight soon passes into complete darkness of the lower bathyal zone.

On the seafloor, the number of suspension feeders (sea pens, colonial octocorals, sponges) decreases rapidly downslope, as the supply of organic detritus from the plankton and the constant wash of currents also diminishes. They are replaced by bottom-scavenging deposit feeders such as the red crab, squat lobster, and sea cucumbers. In very few places this rather monotonous and widely dispersed benthic fauna gives way to an oasis of diversity and profusion. Sponge reefs, for example, present a living prickly carpet of densely packed hexactinellid sponges, harbouring a wide variety of smaller organisms and attracting a range of more mobile predators. Cold, deep-water coral communities are every bit as complex and varied as their warm-water counterparts, but flourish in the bathyal zone without a source of light.

Of the free-swimming fauna encountered at these depths, some are just daytime visitors who rise again at night to feed near the surface in the relative safety of darkness. Others, such as squids, octopuses, sting rays, and deep-sea cod, together with the occasional larger predator such as tuna, sharks, and killer whales, descend to the twilight zone in search of food. Many deep-sea sharks live out their entire lives between twilight and the zone of perpetual darkness below. The largest is the sleeper or Greenland shark, growing up to 7 metres in length and patrolling slowly but constantly over the seafloor. The smallest is the pygmy shark, the size and shape of a fat Cuban cigar, while one of the most fearless and aggressive is the cookie-cutter shark that swims fast and ferociously towards any larger animals, sinking its razor-like teeth into the flesh and taking a circular bite. Many of these deep-water predators are covered with bioluminescent organs, especially over their ventral surface, so they shimmer ghostly green through the darkness.

The true inhabitants of the deeper bathyal and abyssal zones are sparse and peculiarly adapted to their surroundings. With body temperatures close to that of the ambient water, their metabolic rate is very slow. They move and grow more slowly, reproduce less frequently, and live longer than similar species from the surface. Most deep-water species are also smaller, but a few are truly giants. Sea urchins can measure up to 30 centimetres in diameter (five to ten times the size of their shallow-water relatives); hydroids and sea pens off the coast of Japan reach heights of 2.5 metres; while bright red shrimps grow to twenty times the size of those in our prawn cocktails. The most spectacular giant of the ocean floor is the giant squid, averaging from 9 to 16 metres in length. This mollusc is by far the largest known invertebrate, with eyes the size of footballs and tentacles as thick as human thighs, each bearing over 100 suckers with serrated edges.

Perhaps the most remarkable deep-sea habitat, first discovered in the late 1970s, is that surrounding hydrothermal vents that emanate from mid-ocean ridges and spew hot (300–450°C), metal-rich waters directly on to the ocean floor some 2500 metres below the surface. Surrounding the vents for a few metres in all directions are rich and complex marine communities. These include clams, mussels, sea anemones, barnacles, limpets, crabs, worms, shrimps, and fishes—most of which were new to science and are unique to vents. The most impressive of several new species of tubeworm are as thick as a human arm, up to 3 metres long, and have a blood-red gill-like structure protruding from the tips of their white tubes. Pompeii worms are another variety found in cabbage-like clusters closest to the emerging water.

Whatever the vent community, it is chemosynthetic bacteria growing on and within the tissues of many different organisms that are the primary producers. They are capable of oxidizing the normally lethal hydrogen sulphide, thus providing energy to manufacture organic compounds from carbon dioxide. The entire community, therefore, is based on chemical and heat energy derived from within the Earth, rather than on the external solar energy that drives photosynthesis. Some animals feed directly on the bacteria, such as the self-grazing shrimps, clams, and mussels, while others absorb organic molecules released from the bacteria when they die. The tubeworms have symbiotic bacteria making up some 50 per cent of their body weight and so have no need for a mouth, anus, or digestive system. Clams and mussels may comprise as much as 75 per cent bacteria, but they still retain filter-feeding capability and a rudimentary digestive tract. Strange eelpout fishes nibble at the worms and clams.

Chapter 9
Ocean bounty

To the energy sector, the ocean represents oil, gas, and coal, clathrates (frozen methane), and renewables in abundance. To the metals industry, it's manganese, copper, nickel, cobalt, tin, aluminium, molybdenum, titanium, zirconium...as well as gold, diamonds, and precious gems. To fisheries, there's a catch of 100 million tonnes a year—and still growing. To tourism, there are golden beaches and luxury cruises, silent fjords and island tranquillity, miles of hotels, timeshares, and coastal apartments. To the military and to politicians, it represents missile bases and sheltered harbours, trade routes and waste disposal, supremacy at sea and wealth for the future. But to most of us, the oceans will always remain a breath of fresh air and a safe playground for our children.

In a world where the population is set to exceed 10 billion by 2100 and global aspirations are for ever-improved standards of living, it is inevitable that the world's natural resource base is under scrutiny. It is also inevitable that we are turning more and more to the oceans in this context. However, although the ocean is bountiful it is not an inexhaustible solution to our needs. This chapter examines the variety of resources and their availability in the ocean realm.

Metals from the sea

Manganese nodules were first discovered during the pioneering scientific voyage of HMS *Challenger* between 1872 and 1876 (Figure 29). They are concentrated at depths between 4000 and 5000 metres, either closely packed or widely strewn across the seafloor. They are typically potato-shaped, ranging in size from 2 to 15 centimetres, brownish-black in colour, and strangely light in weight—in fact a bit lighter than a potato of similar size.

The nodules are mainly composed of manganese and iron oxides and hydroxides, abundant metals on land and so presently of little economic interest. However, the attractive feature is their high content of copper, nickel, and cobalt, averaging 2–3 per cent combined weight in places. Numerous other metals are scavenged during their growth, including zinc, lead, molybdenum, titanium, and vanadium, which further add to their ore value.

Many scientific questions remain. How, why, and where do they grow? How fast do they grow and are they found buried beneath the sediment surface? We can answer these questions partially. Growth rates are minimal, typically only a few millimetres per million years. This is far slower than the rate of sedimentation over much of the ocean floor, and yet they mostly occur at the sediment surface. The larger nodules must have been growing in the same place for 10–15 million years without becoming buried in sediment—this presents us with an apparent paradox. Three different mechanisms may help explain their position preferentially at the surface: (1) strong bottom currents sweep away any suspended sediment; (2) benthic organisms remove any sediment that does fall; and (3) the metals are continuously dissolved and then reprecipitated on the seafloor.

Two recent discoveries have further added to the inventory of metallic wealth in the oceans. In the early 1960s metal-rich muds

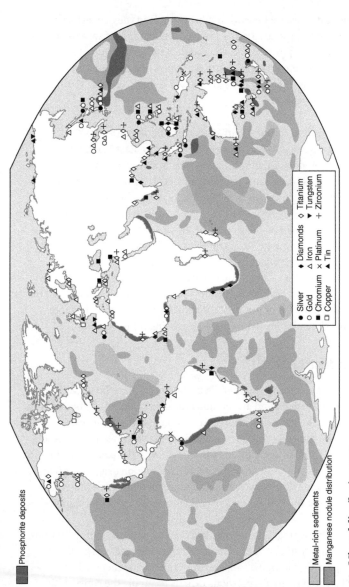

Phosphorite deposits

Silver   ◆ Diamonds   ◇ Titanium
○ Gold   △ Iron   ▽ Tungsten
■ Chromium   × Platinum   + Zirconium
□ Copper   ▲ Tin

Metal-rich sediments
Manganese nodule distribution

29. Mineral distribution map.

were discovered beneath hot brine pools in the deepest parts of the Red Sea. The Atlantis Deep is one example of several narrow elongate deposits, containing some 3 million tonnes of zinc, 1 million tonnes of copper, 80,000 tonnes of lead, and 5000 tonnes of silver. Similar deposits are known from the Gulf of California and Guaymas Basin.

The second discovery, in 1975, was of hot (250–350°C) hydrothermal fluids discharging along the East Pacific Rise, associated with metal-rich deposits. Many such vent sites, known as black smokers, have now been found along the crest of the global mid-ocean ridge system, where seawater circulates through newly formed hot volcanic rocks, scavenging metals into solution and heating up in the process. At some sites, giant tubes or chimneys are formed of condensed metal sulphides (commonly copper, iron, and zinc), in places up to several hundred metres high. Although still some way from economic recovery, preliminary data show that such bodies contain several million tonnes of ore, and compare favourably with the largest massive sulphide deposits that are being mined on land.

Industrial wealth

With an annual output of over 8 billion tonnes, the aggregate business—sand and gravel, concrete and cement—is by both volume and value the second-largest extraction industry in the world after hydrocarbons. Japan is currently the world leader in offshore production of aggregates. However, as with the planned mining for metals on the seafloor, the continued extraction of aggregates must tread a fine line between resource sustainability, environmental degradation, and recovery of a much-needed industrial commodity.

Marine aggregate resources, particularly in those countries that are highly industrialized and densely populated, are gradually replacing traditional supplies from rivers and other onshore areas. Beaches are the first obvious targets to supply the construction industry, but their use now presents a serious conflict of interest with the tourist industry. The next step is offshore, where at least half the area of the broad expanse of continental shelves is covered by sand and gravel. These are relict deposits from former rivers, beaches, and coastal dunes, formed when the sea level was much lower during the past Ice Age.

The other part of the construction equation, a source of lime for cement production, is still very largely available from large *onshore* limestone quarries. But island communities with no suitable supply on land must turn elsewhere. The beach sands of coral islands are composed of broken reef fragments and shell debris, and so present an attractive alternative source of lime. Similar production now occurs off Sri Lanka, Australia, and the United States. Even Iceland, with its interior of dark volcanic rocks and complete absence of limestone, has turned to offshore dredging for shell-rich sand and gravel.

The list of industrial minerals gained from the shallow marine environment continues to increase. Very pure quartz (silica) sand is taken from some beaches directly for the glass industry. A pure form of calcium carbonate (aragonite) is produced in very large quantities from the shallow lagoons of the Bahamas. Locally known as 'white gold', aragonite is an odourless, tasteless, and non-toxic substance with multiple uses in the steel, glass, chemical, agricultural, and food industries. Phosphorite is a sedimentary material containing various phosphate minerals, of prime importance in the manufacture of fertilizer. It is widespread as a recent deposit across continental shelf and upper slope regions in many parts of the world, generally at depths less than 1000 metres—a potential resource for the future.

Salt and freshwater

The oceans are a vast storehouse of dissolved minerals, including all 102 naturally occurring chemical elements. Most are disseminated in such trace amounts that their extraction is quite uneconomic. Very few are sufficiently abundant as well as useful that they are currently extracted on a commercial basis. The most common of these is sodium chloride, or common salt, which makes up 71 per cent of the dissolved solids found in seawater.

Salt has been of prime importance for cooking and for trade over at least the past 5000 years, and has clearly been an item of significant value. The word *salary* derives from the fact that Roman soldiers were paid partly in salt—*salarium argentium*. Then, as now, the principal means of extraction is by direct evaporation of seawater—several stages of crystallization being used to remove some of the magnesium, calcium, and iron compounds that otherwise make for a rather bitter taste. Today, evaporation accounts for about one-third of the world supply of salt, mainly from large-scale plants in India, Mexico, France, Spain, and Italy.

For arid regions, an even more important extract from salt water is fresh water. Desalination plants are a vital way of obtaining scarce supplies for both domestic and industrial purposes. They are becoming increasingly common throughout the Middle East, the Mediterranean, and the North African region, and look to become still more important as an added supply for major coastal cities the world over.

Oil and gas

Although oil has been a part of the human story from the very earliest times, its popular and abundant use has been a remarkably recent affair. By the time Herodotus was writing in 450 BC, pitch and asphalt were being extracted from natural oil seepages all over the Middle East and North Africa—the principal uses being medicinal, waterproofing, warfare, and oil lamps. But, for more than two millennia and right through the industrial revolution, there was little interest in the commodity. Even by the beginning of the 20th century, oil and gas were scarcely used at all. By the beginning of the 21st century, the age of oil had dawned—oil and gas accounted for nearly 60 per cent of commercial energy in the world (Figure 30). We are also living in the age of plastics, and of 1001 other petrochemical products. It would be hard to imagine the world now without low-cost, petrol-fuelled transport, and without plastic.

Oil and gas, together with coal and peat, are fossil fuels based on the principal chemical elements common to all living matter—carbon and hydrogen. They are formed from the organic remains of dead plant material incorporated into sediment, through a process of decay and change over millions of years. Land plants tend to produce gas, while marine phytoplankton yields mainly oil, followed by gas at higher temperatures. The

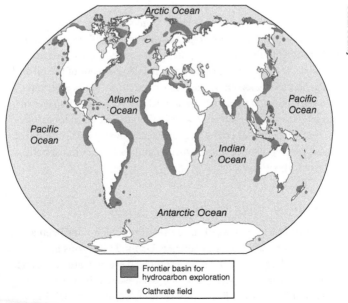

30. Oil frontiers distribution.

process by which organic matter matures is immeasurably slow, and depends upon the increase in temperature as it is buried beneath more and more sediment.

There is a particular window of temperature, between about 50 and 100 degrees Celsius, through which the very complex organic molecules are broken down into the long-chain hydrocarbons found in oil. At higher temperatures still (150–250°C), these are further cracked to yield the short-chain hydrocarbons, mainly methane, of natural gas. The typical thermal gradient, or rate of increase in temperature, is around 20 to 30 degrees Celsius per kilometre burial depth. This means that oil generation commonly occurs between 2 and 4 kilometres below the seafloor, and gas at greater depths. It may take 10–20 million years of sediment accumulation before burial to these depths is achieved.

The very great pressures that result from such deep burial force any mobile hydrocarbons, together with water still trapped in the sediment from the time of deposition, to migrate outwards and upwards. It is estimated that as much as 90 per cent of the oil and gas generated leaks out at the surface in natural seeps and vents. The remaining 10 per cent is held back in the tiny pore spaces of sandstone and other sediments, or in minute cavities in limestones, where shell debris and coral fragments have been partially dissolved. These reservoir rocks fill up with oil and gas rather like a bath sponge fills with water.

Triumph of engineering

Exploration and exploitation of the marine realm has been a triumph of technology and science. Drilling platforms are among the largest structures ever built. They are floated or carried by even larger transporter ships to wherever in the world they are needed, and then fixed in position for months of drilling or many years of production. Accommodation may be required for up to 300 workers at a time, personnel transport by helicopter or boat, and a never-ending round of supplies brought in by ship.

The drilling process is even more remarkable, especially in the deep oceans. Some 200 ten-metre-long steel drill pipes must be carefully joined one by one on the drilling platform to reach a seafloor depth of 2000 metres. Reservoir targets may then be as much as 5000 metres below the seafloor. The holes drilled are like a collapsible telescope of decreasing diameter, from nearly 1 metre at the surface to only 10 centimetres at the base. They are drilled in stages, tested, and then cased in concrete to avoid hole collapse before drilling deeper. Giant steel-knobbed drill bits are used to chew through the softer upper layers of sediment. Fine diamond-studded bits are needed when the sediment becomes hard rock. At all times the advancing hole is cooled, lubricated, and weighted with a dense mixture of mud and water. Without this precaution, the pressure released on piercing a deeply buried reservoir would cause a blow-out with devastating consequences for those above. Commercial drilling routinely takes place in water depths in excess of 2000 metres, and anywhere from the ice-covered Arctic Ocean—drilling from artificial islands—to the sweltering tropics.

Although oil and gas are such an indispensable part of today's world, they are ultimately non-renewable resources that require careful management. Their use contributes to global warming, ocean acidification, and marine pollution.

Frozen gas beneath the seafloor

A new submarine potential lies in the discovery of frozen methane gas at shallow depths below the seafloor. Vast fields of mixed gas and ice crystals, known as gas hydrates or clathrates, are now known to cover millions of square kilometres only a few metres below the seabed. The gas is formed during the very early stages of organic matter degradation, rather than by the thermogenic process described earlier for oil and natural gas, and then freezes under conditions of high pressure and low temperature at water depths in excess of about 1000 metres.

Where gas hydrates have been penetrated by drilling and brought to the surface in cores, the ice melts and gas escapes as pressure is released. Physically, they appear as a white powdery snow intermeshed with sediment, and occur in at least two forms. The first has up to eight methane molecules within forty-six water molecules, and can contain other gases such as ethane, hydrogen sulphide, and carbon dioxide. The second has a larger network of 136 water molecules and can contain higher hydrocarbon gases—pentane and butane.

Initial estimates suggest that their potential as a future energy resource dwarfs existing stocks of other fossil fuels, although the technology and economic feasibility for their exploitation is not yet in place. However, we must be extremely cautious about unbridled exploitation. They occur at very shallow depths within the sediment and any operational accident could release to the oceans and atmosphere one of the most potent greenhouse gases.

Renewable marine energy

On a small and local scale, the use of marine energy is as old as human interaction with the sea. The first fishing rafts used sea currents to drift between islands, and all early voyages of discovery or trade were powered by wind-filled sails. Along the coastline, we soon developed the means to harness such energy more systematically—tidal mills and windmills became commonplace as human population and industry both expanded.

But each year the Sun supplies Earth with nearly a million times as much energy as is locked up in all the planet's oil reserves. A significant proportion of this solar energy is transferred to the oceans and winds. The modern approach, therefore, is to think big. This means building a major dam or barrage across an estuary having a large tidal range—say 4–12 metres. The high tide is temporarily held up in an enclosed bay behind the dam, and then let out through a series of large turbines, generating electricity directly. In fact, the generators can be used in either direction, yielding power from both ebb and flood tides.

The first large-scale commercial scheme was the Rance barrage near St Malo in France, a 1-kilometre-wide dam with twenty-four turbine generators, across an estuary in which the mean tidal range is 8.17 metres, peaking at 13.5 metres during equinoxes. This has been feeding over half a million kilowatts into the French national grid system on each tide since it opened in 1967. Surprisingly, few other tidal barrages yet exist—these being in Norway, Canada, China, and the former Soviet Union—although detailed plans and costings have often been drawn up. The current proposal for the Severn tidal barrage in south-west England would have 200 turbines generating more than 7000 megawatts of electrical energy, roughly equivalent to the supply from five nuclear reactors.

Waves are another obvious and very visible source of marine energy. The power potential of an average wave per kilometre of beach has been estimated at about 40 megawatts, but the means of harnessing this potential is largely in the research and development phase. A small-scale but commercial wave power plant was constructed off northern Scotland in 1995. Wave motion entered into a submerged chamber open at the base, forcing air upwards through turbines. The 2 megawatt facility fed the electricity generated via an underwater cable to the shore some 300 metres away. Unfortunately, the plant was soon damaged by waves and later destroyed by a storm.

Still more elusive is the power held within an ocean current, but this is far from realizable with present technology. However, utilization of the thermal energy in the oceans and converting this directly into electricity, a process known as ocean thermal-energy conversion (OTEC), is more advanced. This uses the significant difference in temperature (around 25°C) between surface and deep water. The warm surface water is used to vaporize an intermediate fuel, such as ammonia, which in turn powers the turbine. Cold water pumped up from depths is used to condense the ammonia back into liquid form, thereby creating a vacuum that causes the gas to circulate. Although pilot schemes have been developed, they are still energy inefficient. But, it has been calculated that the energy locked up in such thermal gradients within the oceans is nearly 10,000 times that available from tides and waves combined.

The use of winds at sea has just begun. Nearly one hundred countries now have some form of wind power, but generally local and small-scale, and almost all onshore. Stronger and more consistent winds are found at sea. This fact, together with issues of competition for land utilization and the visual intrusion of onshore wind farms, is now beginning to turn attention towards offshore potential. Both the UK and the Republic of Ireland have offshore developments, but Denmark is clearly leading the way. There are now twelve fully operational offshore wind farms contributing a significant proportion of the country's electricity supply. Already over 40 per cent of Danish electricity is from combined onshore and offshore wind.

The living resource

The ocean has always been a source of food for humans. Our early ancestors gathered seaweed and shellfish along its shores, caught fish, and hunted seals. Before long they were fishing from boats and had even begun freshwater aquaculture. Through the past several millennia a maritime way of life grew up around small-scale subsistence fishing for countless coastal communities.

Today, seafood provides the principal source of protein for more than 1 billion of the world's people. Fish are also rich in natural oils and vitamins D and B. For some developing countries like Bangladesh, fish provides 80 per cent of the people's needs, while those living in small island states may survive almost wholly on protein from the sea. In such countries fishing is still mainly a small-scale affair, for subsistence and for the local market, which rarely poses a threat to fish stocks. Despite the large numbers fed, these operations collectively account for only about 10 per cent of the global catch.

Large-scale commercial fishing now accounts for 90 per cent of the global catch and is currently running at around 100 million tonnes a year (Figure 31, Table 8). Just over two-thirds of these fish and shellfish are sold directly for food, while the rest is ground up to make fish meal and oils. These are used in animal feed, for poultry, livestock, and mariculture, as fertilizer, and in the manufacture of soap, glue, and paint. A substantial but unrecorded amount, perhaps 20–30 per cent of the total, is caught as by-catch and subsequently discarded at sea. These are either unwanted species or too small to be commercially viable.

The top producing nations, operating throughout the year in every ocean, are China and Peru, followed by Japan, America, Chile, Indonesia, Russia, and India. Fishing fleets may number several hundred vessels with one or more gigantic factory ships. They fish with drift nets that stretch over tens of kilometres, giant purse seines, and open-mouthed trawl nets that can scoop up 90 metric tonnes of bottom-dwelling fish and shellfish in a single trawl. Because the fleet will often stay at sea for many months, the catch must be processed and packaged on board. Just one factory ship is capable of salting 200 tonnes of herring, processing 150 tonnes of mixed fishes to fish meal, filleting and freezing 100 tonnes of flat fish, and manufacturing 5 tonnes of fish oil, all in a single day.

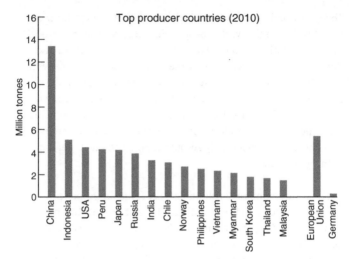

31. Global fish catch.

Half the world's annual catch is taken from the North Atlantic, North Pacific, and the west coast of South America. In these regions, upwelling currents bring cool, nutrient-rich water to the surface, encouraging the rapid growth of phytoplankton and zooplankton during the spring and summer months. This attracts a variety of small crustaceans and fishes that feed and breed in abundance. It is the larger fishes that feed on these animals which are most often the target for commercial fishing. Several hundred different species of fish and shellfish are harvested from the sea worldwide, although relatively fewer types dominate the commercial catch. These depend on the fishing grounds that are active, the availability of particular stocks, and on market demand. Fishing fleets have the latest technology available. Satellite remote sensing can provide immediate data on such parameters as phytoplankton abundance and hence specific information on the most likely feeding grounds that day. Sonar equipment is then deployed to pinpoint large schools of fish hiding below the surface.

Table 8 Fish production statistics (2014)

Wild fisheries world capture production in tonnes

Freshwater fish	10,590,000
Diadromous fish	1,724,800
Marine fish	65,950,700
Crustaceans	6,870,200
Molluscs	7,674,300
Other types	1,382,800
TOTAL	**94,182,800**

World fisheries aquaculture production in tonnes

Fish farms (all)	49,862,000
Crustaceans	6,915,100
Molluscs	16,113,200
TOTAL	**72,890,300**

World aquatic plant production in tonnes

Seaweed aquaculture	27,307,000
Seaweed wild	1,184,500
TOTAL	**28,491,500**

Source: FAO.FAOSTAT <http://www.fao.org/fishery/topic/16140/en>.

However, marine ecologists generally agree that commercial fishing is the single greatest pressure on marine ecosystems. Serial overfishing of popular species has resulted from improved technologies, larger fleet sizes, and increased demand. Stock depletion of larger fish, such as cod and haddock in the 1970s and 1980s, led to severe depletion and the need for regional fishing bans in order that populations might recover. Introduction of the purse seine net in the 1960s directly led to the overfishing of herring stocks. As stock depletion affects the higher trophic levels, the industry turns its attention to smaller species, including anchovies, sardines, pilchards, and capelin. This trend of fishing down the food chain is recognized as unsustainable.

The blue revolution

By analogy with the 1960s green revolution in land-based farming, which greatly increased agricultural production in developing countries, the recent upsurge in mariculture (marine fish farming) has been heralded as the *blue revolution*. In reality, the Chinese have been farming fish, shellfish, and seaweeds for some 4000 years, and even wrote the first textbook on the subject in 475 BC. This described the farming of freshwater carp, brackish water milkfish, and marine mullet. The practice has continued, especially in South and East Asia, where shallow ponds are dug in coastal areas so they are readily replenished by the incoming tides. These ponds are seeded with the eggs of fish and crustaceans, and the larvae or juveniles then transferred to larger ponds where they are fed on larvae and plankton. Oysters and mussels also have a long history of cultivation in many parts of the world.

World mariculture production is currently around 15 million tonnes, and has become commercially more important than freshwater aquaculture, especially in China and Japan. Oysters, scallops, carpet shells, and molluscs are produced in vast quantities, as are prawns, shrimps, and lobsters. Salmon, amberjacks, and freshwater carps are the principal fishes, followed by Japanese yellowtails, Florida pompano, and a variety of flatfishes. Exotic items, such as giant tiger prawns, Pacific cupped oysters, and sea-urchin roes, are the most valuable in monetary terms. Seaweed harvests yield the greatest tonnage, not only for consumption as food, but also to create the smooth texture of ice-cream, toothpaste, and paint, and to increase the clarity of beer.

But, as for the green revolution before, the blue revolution is not without its problems and critics. Intensive farming also leads to the potential for the rapid spread of disease, such as fish lice commonly found on salmon farms, and this may even spread outwards to infect wild fish stocks. There is debate about the rights and wrongs of genetic engineering, to enhance weight gain, improve flavours, and increase resistance to disease, for example. The use of antibiotic supplements in fish foods may counter the problem of disease but creates an unwanted surfeit of antibiotics in the wider environment and builds resistance in bacteria. Ecologically, it is often true that the production of fish food for mariculture consumes more biomass (small crustaceans and fishes turned into fish meal and oil) than is produced in the form of prawns, salmon, and other table fishes.

Chapter 10
Fragile environment

As we turn progressively to the oceans for their natural resources, as well as for trade, pleasure, and tourism, we need to be fully aware that resources have different limits to their renewability, and that marine ecosystems are profoundly fragile. The 10 billion people we expect to populate the planet by the end of this century all need adequate and sustainable energy, drinking water, and food supplies. These are three of the most pressing issues of the 21st century and for each of them the oceans hold part of the solution. I would further argue that sustainability can only be achieved through a more equitable distribution of resources, a limit to excess, and firm management of the environment.

The new global population is also set to produce about twice the amount of waste that we do today. The seas are neither so large nor marine ecosystems so robust that they can withstand these pressures without change. The coastal regions are already home to more than 3 billion people and that figure is set to double by 2020. Marine tourism is a booming multinational industry. The pressures on our coastline from urban sprawl, industrial development, and recreational demand are without precedent. Climate change will influence the oceans in ways that have knock-on effects for its many dynamic balances—ocean acidification, nutrient cycles, carbon storage, and more.

This information and the caution it requires by humankind is perhaps the single most important message that ocean scientists and a supporting public need to convey to our politicians and planners. This concluding chapter, therefore, takes a closer look at stresses on the ocean environment and considers the options and challenges we face for achieving sound management of ocean space in the coming decades. Although the statistics presented may appear dramatic, they are simply data that illustrate the scale of the problem.

As an ocean scientist myself, I have always been aware that the more we understand the seas and seafloor, the more we can put that knowledge towards its protection and management. Deep-sea oil and gas, and renewable energy from the oceans, both figure prominently. The transition from fossil to sustainable energy has begun and an exponential growth in global consumption of renewable energies is underway. Both have the capacity to affect ocean environments.

Marine pollution

The sheer scale of waste generation across the globe is quite staggering (Figure 32, Table 9). The world population is now in excess of 7 billion and appears set to reach 10–11 billion by the year 2100. Collectively we produce over 7 billion tonnes of domestic waste each year, unevenly distributed with a strong positive correlation between income level and waste generation. Industrial and agricultural operations, engineering works, and energy production yield waste products in equally large amounts. There is a further 12 billion tonnes of sewage sludge per annum. Currently, a large proportion of these various sources is disposed of or treated on land, but marine dumping is on the increase, particularly along the coastal zone in highly populated regions of the world. Over 4 million tonnes of oil is spilt into the sea every year, focused along shipping lanes and coastal regions.

Oceans

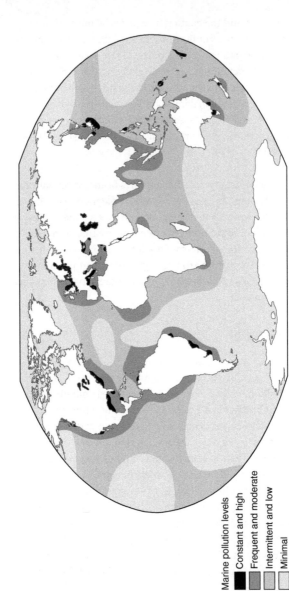

Marine pollution levels

■ Constant and high
▨ Frequent and moderate
▧ Intermittent and low
□ Minimal

32. Marine pollution levels map.

Table 9 Principal pollutants

Principal pollution types and sources	Type	Examples	Human sources	Effects
	Nutrients	Nitrogen, phosphorus	Sewage outflow, agricultural runoff, air-borne oxides from burning fossil fuels	Proliferation of algae leading to oxygen depletion and toxicity, harming marine life
	Pathogens (bacteria, viruses, parasites)	Hepatitis, cholera, typhoid, diphtheria	Sewage, livestock	Contamination of beaches and shellfish, leading to spread of disease
	Sediments		Erosion from mining, forestry, farming, etc. Also, dredging, water engineering	Clouding of water, leading to reduced photosynthesis and smothering of ecosystems
	Toxins (heavy metals, organohalogens)	Mercury, lead, cadmium, arsenic, DDT, PCBs	Industrial effluents, pesticides, urban wastewater discharge	Poisons coastal marine life; accumulates in food chain to cause disease and disrupt reproduction in top predators

(continued)

Table 9 Continued

Principal pollution types and sources	Type	Examples	Human sources	Effects
	Hydrocarbons	Petroleum (crude oil)	Urban runoff, oil transport and shipping, refineries, atmospheric fallout, tanker accidents, offshore production	Oil spills in coastal areas devastate local biota; prolonged low-level contamination can cause disease and reduce productivity
	Radionuclides	Uranium, plutonium, caesium	Discharges from power stations, atmospheric fallout, discarded military waste	Contamination of shellfish leading to diseases such as leukaemia in humans
	Litter	Plastic bottles, fishing gear, disposable nappies, aluminium cans	Fishing, shipping, tourism, industrial and landfill wastes	Animals ingest plastic items or become entangled; debris can remain on beaches for 400 years

At least 150,000 tonnes of plastic nets and other fishing gear are abandoned at sea annually, and plastic refuse can be found everywhere.

There are a number of waste products that are naturally and quickly removed by the marine environment; others remain for much longer. Organic substances, such as sewage sludge, are biodegraded and recycled within a few days to a few months provided that the environment is not unduly overloaded. Oil and oil products are effectively removed when present in small amounts, although large spills and shipping-lane pollution will stress the natural coping mechanisms. Thermal pollution from industrial emissions and cooling plants is also rapidly dissipated.

Many artificial products, however, are longer lived than natural substances and more dangerous to the marine environment. Being synthetic and only recently introduced into the natural world, they are more resistant to degradation by bacteria and other micro-organisms. Detergents, pesticides, and other organo-chemicals are very persistent over periods of years to many decades, while radioactive isotopes and toxic metals can last for centuries and longer.

Two especially long-lived organo-halogens were introduced in the 1950s and 1960s and much heralded at the time for their beneficial properties. These were DDT (dichloro-diphenyl-trichloroethane), the chief component of agricultural pesticides, and PCBs (polychlorinated biphenyls), used in paints, plastics, adhesives, hydraulic fluids, electrical appliances, and aerosols. They have found their way into rivers and seas throughout the world, where they are incorporated into the fatty tissues of living organisms and become concentrated along the food chain as one animal eats another. They act to inhibit normal growth and cause infertility in many vertebrates. Recognition of their severely toxic effects has now led to a major reduction in their use.

The sewage debate

Sewage comprises organic and nitrogen compounds, ammonia, and concentrations of heavy metals. Because it is readily biodegraded, it has commonly been pumped directly out to sea. Major urban populations of North America (New York, Los Angeles, and Halifax, for example) have routinely dumped untreated sewage into their coastal waters over the past half century. The same has been true for north-west European cities surrounding the North Sea, and for the growing population that fringes the Mediterranean. Sewage input into this inland sea is especially heightened during the tourist season.

The consequence of overloading the marine environment with sewage can be quite dramatic. The influx of abundant nutrient elements creates a flourishing micro-cosmos of competing organisms (viruses, bacteria, and other microbes) and consequent rapid reduction in oxygen levels. Water anoxia occurs, resulting in reduced biotic diversity, shellfish poisoning, and an increased likelihood of toxic phytoplankton blooms (red tides). These release specific toxins into the water causing the death of marine mammals, fish, and invertebrates, as well as the emergence of certain common disease pathogens. Hepatitis, typhoid, dysentery, and enteritis are all associated with sewage contamination, and a new form of cholera has recently appeared in several Asian nations. Although the recent instigation of minimum standards and controls throughout Europe and much of North America requires better sewage treatment and the progressive cessation of most marine dumping, this is not yet true everywhere in the world.

Oil pollution

Natural seepage of oil and gas into the marine environment is estimated at around 600,000 tonnes per year, which is widely dispersed and readily biodegradable. A similar amount is added to the oceans through spillage associated with its production and usage. Although the spillage of less than 1 million tonnes a year into the vast ocean reservoir is only about 0.01 per cent of the 6 billion tonnes that we consume each year, its concentration along shipping routes and coastlines, in harbours and areas of major offshore oil production, is the principal cause for concern. Such long-term, low-level contamination can kill larvae in coastal nurseries, cause a variety of diseases among marine organisms, and generally lead to negative changes in population structure and dynamics.

The majority of oil spillage results from routine operations, such as loading, discharging, and bunkering, coupled with runoff from cars, heavy machinery, industry, and other land-based sources. Shipping accidents and collisions are very intermittent in occurrence, although the scale of spillage can be very large, especially from supertankers, and with huge environmental consequences for the regions affected. The largest ever was due to the collision between the *Atlantic Empress* and *Aegean Captain* off Trinidad and Tobago in 1979, when 270,000 tonnes of oil emptied into the Caribbean. The *Exxon Valdez* disaster in 1989, near the Valdez oil terminus in southern Alaska, led to the loss of 35,000 tonnes at sea and devastation to over 2400 kilometres of pristine coastline north of the Arctic Circle.

Rarer still but potentially worse are accidents to oil wells at sea. Blow-outs occur when a highly pressured reservoir of oil several thousands of metres below the seafloor is penetrated with insufficient weight of drilling mud to keep the oil in place. Two memorable disasters of this kind both occurred in the Gulf of Mexico. The *Deepwater Horizon* explosion in 2010 led to an estimated 675,000 tonnes of oil entering the ocean over the four-month period it took to bring the leakage under control. The second largest occurred beneath the drill rig *Ixtoc I* in 1979 and was not brought under control for almost eleven months. During this period nearly 535,000 tonnes of oil escaped.

The worst single oil disaster ever was a deliberate act of ecological terrorism by Iraq's former President Saddam Hussein as his troops retreated from Kuwait in 1991. More than 600 well heads were set alight—black smoke filling the air for many months, and thick layers of soot were found more than 3200 kilometres away in the Himalayas. At the same time, oil refineries along the coast were sabotaged, creating one of the biggest oil slicks known to civilization and causing catastrophic environmental damage in a region of the world already under severe stress from chronic oil pollution.

While the immediate effects of major disasters such as these can be severe, the long-term effects on marine ecosystems are still poorly understood. In regions of persistent oil pollution, including many inland and marginal seas as well as the coastlines adjoining heavily populated or industrialized areas, marine life is permanently affected.

Habitat destruction

Concern for the environment is a relatively recent concept, so that mismanagement has already blighted great tracts of the world's land area. Continental habitats have often suffered irreparably: marching desertification and deforestation; urban and industrial sprawl; soil erosion and biodiversity loss; and are now facing the little-known effects of global warming. So too are the ocean habitats under threat as never before. Although the oceans are vast, they are not so vast that they can remain untouched by human activity.

The coastal zone is the first to be affected, including a wide variety of habitats—the rocky shoreline and sandy beaches, barrier islands and sheltered lagoons, muddy estuaries and delta plains, salt marshes and seagrass beds, coral reefs and mangrove swamps. Together, these regions boast a biodiversity that is second to none, and marine productivity of the first order. About 90 per cent of the world's total seawater fish catch reproduces in coastal areas. About half the world's population—over 3.5 billion people—lives close to the sea, and two-thirds of all larger cities, with populations in excess of 2.5 million, have been constructed near estuaries or deltas. Continued migration to these cities will see the population of the coastal zone increase by as much as 50 per cent in the next two decades.

It is the coastal environments, therefore, that are most fragile and most affected by human activities. A combination of factors contributes to the serious problem of sinking cities around the world: excessive pumping of groundwater from beneath the cities for drinking water; increased rates of sediment compaction due to the sheer weight of infrastructure and removal of water; and rising sea levels due to global warming. Beijing and Bangkok are each subsiding by 11–12 centimetres per year and Djakarta by a massive 28 centimetres per year. Shanghai, New York, New Orleans, and Venice are all still sinking although the rates have been brought under greater control in recent years. There are many other examples.

Seasonal excesses due to the influx of tourists place a particular environmental strain in coastal regions. Tourism remains the world's fastest-growing business and collectively the largest employer with an annual turnover in excess of $4 trillion. It is estimated to generate, directly or indirectly, nearly 250 million jobs, or around 10 per cent of global employment and 30 per cent of global trade. A high proportion of those jobs and of the estimated 1 billion tourists annually is focused on or close to the sea.

Two of the marine habitats that are suffering most ecologically are mangrove wetlands and coral reef complexes. The mangrove ecosystem of salt-adapted trees and spreading roots is one of the most fecund in the world. It provides vital spawning grounds and nursery areas for fish and shellfish, and harbours a multitude of adult organisms and seabirds. It is also vital for the regional fishing industry and livelihoods of millions of people. More than 35 per cent of the world's mangrove forests are already gone; the figure is over 50 per cent in parts of South and South East Asia. The principal causes of this decline include clearance for agriculture, urban and industrial uses, overharvesting for firewood and charcoal, pollution from fertilizers and pesticides, and rising sea levels.

Although coral reef complexes cover only 0.2 per cent of the ocean floor, they support around 25 per cent of known marine species. Some of the richest reefs in the world contain as many as 700 species of corals, 2000 species of fish, and 5000 species of molluscs, as well as numerous other organisms. Reefs act as an energy baffle for waves and storms, protect the coastline from erosion, are vital to both the local and global fishing industry, and play an important role in many island subsistence economies including tourism.

But coral reefs are being degraded at unprecedented rates. Significant reef damage has occurred in ninety-three of the 109 countries with major coral habitats. Over 25 per cent of the world's reefs have already gone and a further 50 per cent are at serious risk. The principal causes are human-induced: land reclamation for building towns, harbours, airports, and hotels in close proximity to the reef complex; extensive mining of the reef to provide construction materials; and major silting-up of the coastal zone due to land clearance and deforestation in the hinterland. But perhaps the most serious threat now is the steady acidification of ocean waters that results from global warming and the uptake by seawater of excess carbon dioxide in the atmosphere.

Bottom trawling as a fishing technique has had a profound effect on seafloor habitats. There is extensive collateral ecosystem damage inflicted by this practice right across the continental shelves, as well as on offshore banks and seamounts in deeper water. Trawling also resuspends sediments that can smother a far wider area than that directly impacted by the passage of the trawl.

Dying seas and virgin habitats

Enclosed and semi-enclosed seas marginal to the major ocean basins have experienced some of the highest levels of pollution, especially where they are surrounded by busy towns, cities, and industry, or where they have become centres of tourism. The whole suite of marine habitats they comprise are gradually put under greater and greater strain, the seas slowly poisoned around the edges, and marine life permanently affected.

Isolated inland seas, such as the Great Lakes of North America and the Caspian Sea in central Asia, have now become completely polluted and are more or less dead. The Gulf of Saint Lawrence, the North Sea and Baltic Sea in Europe, the Gulf of Arabia, and the Yellow and East China Seas, all contain some or many habitats that have become severely polluted and destroyed through long years of industrial, agricultural, and urban activity. The Black Sea, the Mediterranean Sea, and parts of the Caribbean have been similarly affected, now also by a huge seasonal influx of tourists.

Large amounts of nitrogen and phosphorus pollutants, especially from sewage discharges, cause sudden growth explosions of phytoplankton, quickly followed by zooplankton communities. Many of these plankton blooms are quite harmless but others, known as red tides, produce neurotoxins that affect marine life higher up the food chain. In 1988 a 10-metre-thick layer of toxic plankton extended along the narrow straits between the North and Baltic Seas, and led to the deaths of millions of fish. This incident is known as the 'marine Chernobyl', but unfortunately such blooms are now quite frequent.

Although some parts of the open ocean and deep-water habitats remain largely unaffected by human activity, others do not.

Drilling for oil and gas already takes place down to a water depth of 3000 metres. Metal mining operations are likely to follow soon. Marine pollution is most noticeable on the high seas along busy shipping lanes. Even the most remote gyres of the central Pacific and Atlantic Oceans now sport vast islands of plastic waste. Elsewhere, there are hundreds of cold-water coral communities living in deep water, deriving food and energy from surface productivity. There are unique chemical-fed ecosystems surrounding the hot-water vents and black-smoker chimneys discovered along the mid-ocean ridges. There are equally harsh but flourishing habitats in polar seas that are still subject to strict environmental regulation. And there are vast tracts of the deep oceans and abyssal trenches that have never been surveyed. Marine scientists know very little or nothing about these and many other near-virgin ecosystems. We cannot therefore know to what extent they have been or might be affected by our use of the oceans.

Towards a solution

The destruction of natural habitats is not, of course, a uniquely human achievement—the geological past has witnessed much greater changes than those we see today. But since the advent of humans on Earth, and mainly through the last few centuries of history, the style and pace of change has been forced by our own actions. The oceans, like all parts of the Earth system, are a fragile environment, some parts of which have already been severely affected. But they are also a robust system that can cope with change and mitigate its effects.

While ocean scientists learn more about the oceans and their capacity to absorb change, it is wise to take precautionary measures. Fishing quotas and temporary bans are agreed both regionally and internationally, although they are not always easy to enforce. A complete moratorium on commercial whaling was introduced in 1986 with considerable success. Some nations—Japan, Russia, Norway, and Iceland—still refuse to comply and set their own

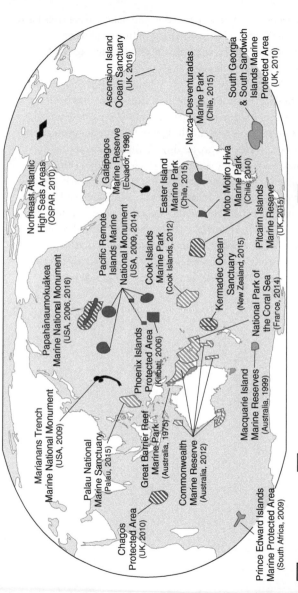

33. **Marine reserves.**

Marianans Trench
Marine National Monument
(USA, 2009)

Northeast Atlantic
High Seas Areas
(OSPAR, 2010)

Ascension Island
Ocean Sanctuary
(UK, 2016)

Papahānaumokuākea
Marine National Monument
(USA, 2006, 2016)

Pacific Remote
Islands Marine
National Monument
(USA, 2009, 2014)

Galapagos
Marine Reserve
(Ecuador, 1998)

Nazca-Desventuradas
Marine Park
(Chile, 2015)

South Georgia
& South Sandwich
Islands Marine
Protected Area
(UK, 2010)

Palau National
Marine Sanctuary
(Palau, 2015)

Cook Islands
Marine Park
(Cook Islands, 2012)

Easter Island
Marine Park
(Chile, 2015)

Moto Motiro Hiva
Marine Park
(Chile, 2010)

Phoenix Islands
Protected Area
(Kiribati, 2006)

Kermadec Ocean
Sanctuary
(New Zealand, 2015)

Pitcairn Islands
Marine Reserve
(UK, 2015)

Great Barrier Reef
Marine Park
(Australia, 1975)

National Park of
the Coral Sea
(France, 2014)

Commonwealth
Marine Reserve
(Australia, 2012)

Macquarie Island
Marine Reserves
(Australia, 1999)

Chagos
Protected Area
(UK, 2010)

Prince Edward Islands
Marine Protected Area
(South Africa, 2009)

Marine Reserve

Marine Reserve*

Multi-zoned (with some No-take) Marine Protected Area

Multi-zoned (with some No-take) Marine Protected Area*

Other Marine Protected Area

* Officially designated, not implemented

quotas. Strict environmental protection and pollution controls have been introduced, but only apply to parts of the world where contiguous nations have recognized the problem and agreed to collaborate towards a solution.

An important development in the past thirty years has been the establishment of marine protected areas (MPAs). These are zones of the seas and oceans that are designated for special protection, especially with a view to preserving biodiversity and the management of fish populations and other resources (Figure 33). At the time of writing, there are over 6500 MPAs covering nearly 3 per cent of ocean space. Most are set up and managed locally and nationally, with different controls and expectations in place. Taken together they are an extremely effective method of marine resource and environmental management. However, the trend must continue and their coverage increase significantly in order to realize their full potential for managing the global ocean environment.

Epilogue: new global awareness

It is only in the past 200 years that we have developed an unprecedented capacity to effect global change. Human action is the principal cause of rapid global warming, progressive acidification of the oceans, the melting of polar ice, and the steady rise of global sea level. All sorts of physical and biological effects derive from these planetary changes, including the destruction of natural habitats and dramatic increase in the rate of species extinctions, the warming of oceans and an increase in the number of tropical storms, and adverse changes to the fishing and mariculture industries.

But the past three-quarters of a century have also seen the rise of a new environmental awareness, expressed by both individual concern and international cooperation. The United Nations Environment Programme was founded in 1945, closely followed by new governmental departments of the environment around the world. Concerned non-governmental organizations also proliferated—Greenpeace, for example, was first launched in 1973. The Law of the Sea Treaty was ratified by the United Nations in 1982 and currently has 167 signatories. Since the first Earth Summit held in Rio de Janeiro in 1992, there have been a series of follow-up summits and important progress made, with new global conventions on biodiversity and climate change.

There is little doubt that sustainable development for the Earth depends on significant advancement in our management of the oceans. Already, we are looking to the oceans to solve some of the major problems we are facing in the world. The blue revolution in mariculture will compensate for dwindling fish stocks, desalination plants will provide universal safe drinking water, deep-water oil and gas can provide a short-term solution to the fossil fuel crisis, while marine renewable energy is a longer-term alternative, offshore sand and gravel resources are believed to be more than enough to construct the megacities and tourist needs of the coming century, and so on. Seafloor sediments might form safe disposal sites for toxic and nuclear waste; while the subsurface can provide repositories for surplus carbon dioxide in our battle against global warming.

But systematic scientific exploration of the seas is little more than two centuries old, and current understanding still in its infancy. The truth is that we do not yet know enough to turn such aspirations into reality. We are toying with the unknown, anticipating ocean properties or ocean responses before we have reliable models to predict the outcomes of our actions. Human need and human greed are quickly outstripping the pace of understanding and back-staging the advancement of scientific knowledge. The concept of ecological unity in the earth–ocean system is fast being replaced by economic imperative.

The use and abuse of the oceans will undoubtedly affect every individual on the planet, so it is far too important an issue to leave in the hands of politicians alone. Already, ocean-related issues are becoming ever more prevalent in the global media—warming seas and the El Niño effect, ocean acidification and coral bleaching, melting of polar ice and sea-level rise, increased storminess at sea, coastal pollution and dying coral reefs, and so on. It is therefore a vital challenge for ocean scientists to ensure that the facts and issues are correctly reported, and that politicians as well as ordinary people have the knowledge base in order to enable sensible decision-making. Developing an ocean awareness and keen global sensibility is the challenge for us all as world citizens.

Glossary

abyssal plain the flat, sediment-covered area forming the greater part of the deep-sea floor, between the continental rise and the mid-ocean ridge, at a depth of 3000–6000 metres.

abyssal zone a subdivision of the benthic ocean province between 4000 and 6000 metres.

acidification become weakly acidic due to the presence of dissolved compounds such as sulphur dioxide and carbon dioxide.

algae simple aquatic plants, unicellular and multicellular, that lack roots, stems, and leaves, e.g. diatoms and seaweeds.

anoxic the absence of free oxygen (O_2).

aphotic zone the dark region of the ocean lying beneath the surface sunlit waters.

archaea one of the oldest forms of life on Earth, found abundantly in extreme environments; single-celled prokaryote; one of the three principal domains or six kingdoms of life.

asthenosphere the hot, soft region of upper mantle that lies directly below the lithosphere.

atmosphere the vast and complex system of gases and suspended particles that encircle the Earth.

autotroph an organism that is capable of producing its own food, such as plants and other photosynthesizers.

basalt a dark, fine-grained, volcanic igneous rock composed of minerals enriched in ferromagnesian silicates; it typifies the oceanic crust.

bathyal zone the portion of the ocean bottom that extends from the edge of the continental shelf to a depth of 4000 metres.

bathymetry the measurement of depth below sea level in the ocean in order to delineate the submarine topography.

benthic pertaining to the ocean bottom.

biodegradation decomposition by biological agents, especially bacteria and archaea.

biogenic sediment sediments formed from the remains of living organisms, mostly an accumulation of their hard parts.

bioluminescence the production of visible light by organisms.

black smoker deep-sea hydrothermal vent, mainly occurring at mid-ocean ridges.

bottom water a general term applied to dense water masses that sink to the floor of ocean basins.

calcareous composed of calcium carbonate ($CaCO_3$).

carbonate compensation depth the depth in the ocean below which material composed of calcium carbonate is dissolved and therefore does not accumulate on the sea floor.

Cenozoic of, belonging to, or designating the latest era of geological time, from 65 million years ago to the present.

chemogenic sediment sediment that forms from the direct precipitation of chemical compounds from water, or from ionic exchange between existing sediment and water.

circulation circular movement of a substance, especially water in the oceans.

clathrate a substance in which the guest molecules occupy a space within the crystal lattice of another substance; as in methane clathrate, in which water occurs within a frozen lattice of water molecules.

coastal upwelling the upward flow of cold, nutrient-rich water that is usually induced by Ekman transport.

coccolith a microscopic calcitic skeletal platelet that helps protect certain marine phytoplankton; the dominant component of certain limestones, such as chalk.

coccolithophore photosynthetic protist, covered in calcitic plates, belonging to the phylum chrysophyta.

continental drift the movement of continental masses as a result of seafloor spreading.

continental rise gently sloping region towards the deep ocean, lying at the base of the steeper continental slope that surrounds the continents.

continental shelf submerged platform region extending from the edge of a continental landmass; generally less than 200 metres water depth.

continental slope the sloping seafloor of the continental shelf between the shelf edge (around 100–200 metres water depth) and the continental rise or deep seafloor.

convection cell convection current completing a circular motion, distinct from others around it; can refer to slow movement of the heat and fluid portion of the mantle, or to air movement in the atmosphere, for example.

convergence zone where two horizontally flowing water (or air) masses meet, and are forced to change direction, typically flowing vertically.

convergent margin a boundary at which tectonic plates collide, typically resulting in volcanic activity, earthquakes, and mountain-building.

core the innermost region of the Earth, beginning at a depth of around 2900 kilometres; thought to be mainly composed of iron and nickel; divided into outer liquid core and inner solid core.

Coriolis effect apparent deflection of the path of winds and ocean currents (or other free-moving bodies) that results from the rotation of the Earth.

crust the outermost, thinnest, and coolest layer of the Earth; consists mainly of either granite (continental crust) or basalt (oceanic crust); thickness varies from 5 to 70 kilometres.

cyanobacteria a group of photosynthetic prokaryotes that contain chlorophyll and that release oxygen as a by-product of their photosynthesis; formerly referred to as blue-green algae, now simply as 'blue greens'.

deposit feeder animal that feeds on bottom sediments, extracting useable organic material and discarding inorganic debris.

detritus either *organic* matter, such as animal wastes and bits of decaying tissue, or *inorganic* sedimentary material (= sediment).

diatom photosynthetic, unicellular protists that belong to the phylum chrysophyta; possess a glassy covering composed of silica.

dinoflagellate photosynthetic, usually single-celled protists, possessing two flagella, belonging to the phylum pyrrophyta.

ecosystem a discrete ecological unit consisting of organisms and their environment.

El Niño warm surface waters that usually appear around Christmas off the coasts of Ecuador and Peru; every three to five years, approximately, the El Niño effect is intensified, bringing extreme weather conditions to many parts of the globe, especially the tropics.

eukaryote cells possessing a true nucleus and other organelles.

evaporite a type of sediment precipitated from a concentrated aqueous solution, usually by the evaporation of water from a basin with restricted circulation; includes halite, gypsum, and anhydrite.

evolution the process by which populations of organisms change over time.

extinction the disappearance of a species from their entire geographical range.

fault natural break or rupture between rocks along which movement takes place; such movement is one of the principal causes of earthquakes.

filter feeder an organism that filters its food from the water.

food chain a sequence of feeding relationships among a group of organisms that begins with producers and continues in a linear fashion to higher level consumers; generally a simplified picture of reality.

food web a representation of complex feeding networks, comprising many food chains or part chains, which exist in an ecosystem.

foraminifer amoeba-like protozoan (single-celled organism), many of which produce an elaborate shell of calcium carbonate.

fossil fuel fuel resource, including coal, oil, and natural gas, formed from the remains of plants and microorganisms that lived millions of years ago.

global warming warming of the Earth's average global temperature due to increasing concentrations of greenhouse gases.

greenhouse effect the warming of the Earth's atmosphere due to the absorption of infrared terrestrial radiation by greenhouse gases.

greenhouse gas gas in the atmosphere that absorbs infrared terrestrial radiation, thus raising the Earth's average global temperature, including carbon dioxide, water vapour, nitrous oxide, ozone, and CFCs.

Gulf Stream warm ocean current that originates in and around the Caribbean and flows across the North Atlantic to north-west Europe.

gyre a large water circulation system of geostrophic currents rotating clockwise (northern hemisphere) or anticlockwise (southern hemisphere).

habitat the specific place in the environment where a particular plant or animal lives.

hadal zone a subdivision of both the pelagic and benthic zones, deeper than 6000 metres; also known as hadalpelagic in the open ocean.

heavy metal inorganic metallic element with high atomic mass, such as mercury, lead, arsenic, tin, cadmium, cobalt, selenium, zinc, manganese, and copper; these may be highly toxic when concentrated in the environment or in organisms.

herbivore an animal that eats only plants and algae.

heterotroph an organism that relies on other organisms for its food; also known as a consumer.

hot spot localized zone of melting and upwelling in the asthenosphere/lithosphere above which volcanic activity is abundant.

hydrocarbon organic compounds composed of hydrogen, carbon, and oxygen; the main components of petroleum (both oil and gas).

ice age period of time when glaciers dominate the surface of the Earth, and global mean temperatures are lowered.

icehouse effect the cooling of the Earth's average global temperature occurring during an ice age, commonly due to an increase in albedo.

interglacial period period of time during an ice age when temperatures increase leading to glacier retreat.

island arc a chain of volcanic islands associated with oceanic subduction zones, lying on the continent-side of deep-sea trenches; formed by the partial melting of the lithosphere as a plate is subducted.

kelp large benthic species of brown algae.

krill pelagic, shrimp-like creatures that belong to the arthropod order euphausiacia.

lava magma extruded at the Earth's surface; used for both the still-molten form and when hardened into volcanic rock.

lithification the process by which loose, unconsolidated sediment is compacted and cemented into a sedimentary rock.

lithosphere the relatively cool, brittle outer shell of the Earth, including the crust and upper mantle; extends to a depth of around 100 kilometres.

littoral zone a subdivision of the benthic province between the high and low tide marks, equivalent to the intertidal zone.

manganese nodule concretionary lump, typically the shape and size of a potato, containing oxides of iron, manganese, copper, nickel, and various other metals, found scattered over the deep ocean floor.

mangrove a dense growth of salt-tolerant mangrove trees and shrubs in marsh-like shoreline environments of the tropics and subtropics.

mantle the thick layer of the Earth's interior between the crust and the core, composed of ferromagnesian silicates.

marginal sea a semi-enclosed body of water adjacent to and surrounded by continent; floored either by submerged continental crust or oceanic crust, or by a mixture of both.

mass extinction a catastrophic, widespread perturbation where major groups of species become extinct in a relatively short period of geological time.

Mesozoic of, belonging to, or designating the era of time including the Triassic, Jurassic, and Cretaceous periods, between 245 and 65 million years ago.

mid-ocean ridge a long mountain range that forms along cracks in the ocean floor where magma breaks through the Earth's crust; the site of formation of new ocean crust and seafloor spreading as two tectonic plates move apart; also known as a divergent plate margin.

Milankovitch cycle cyclic variations in climate, with regular periodicities of around 20,000, 40,000, and 100,000 years, as a result of irregularities in the Earth's rotation and orbit.

Mohorovičić discontinuity a compositional and density discontinuity marking the interface between the rocks of the crust and the mantle; also known as the 'Moho'.

monsoon a regional-scale wind system that predictably changes direction with the passing of the seasons, mostly in South and South East Asia; summer monsoons are often accompanied by heavy precipitation.

nektonic actively swimming organisms.

neritic one of the oceanic zones, referring to water lying over the continental shelf, to a depth of around 200 metres.

oceanic crust the outermost shell of the Earth, around 5–10 kilometres thick, that underlies oceans, composed of the basic igneous rocks—basalt, dolerite, and gabbro—and commonly overlain by sedimentary layers.

oceanic gateway partial topographic barrier between ocean basins, through which a deeper, narrow zone or channel allows the interchange of water masses.

ophiolite a sequence of igneous rocks, thought to represent a fragment of oceanic lithosphere now emplaced onto the continent, composed of peridotite overlain successively by gabbro, sheeted dolerite dikes, and pillow basalts.

organohalogen complex organic compound that includes one or several of the halogen elements (e.g. chlorine, fluorine, bromine) in its structure.

outgassing the process whereby gases derived from Earth's interior are released into the atmosphere and hydrosphere during volcanic and hydrothermal activity.

ozone layer atmospheric concentration of ozone (a form of oxygen with three atoms per molecule) found at an altitude of 10–50 kilometres above the Earth's surface.

palaeoceanography the study of past oceans, their location, and characteristics.

Paleozoic of, belonging to, or designating the era of time including the Cambrian, Ordovician, Silurian, Devonian, Carboniferous, and Permian periods, between 545 and 245 million years ago.

Pangaea the single super-continent of the late Paleozoic and early Mesozoic Eras that comprised all of the present-day continents.

Panthalassa the late Paleozoic to early Mesozoic worldwide ocean that surrounded the super-continent Pangaea.

pelagic of the ocean; more commonly the open ocean away from the shoreline.

Phanerozoic of, belonging to, or designating the period of time (the eon) including the Paleozoic, Mesozoic, and Cenozoic Eras, extending from 545 million years ago to the present.

photic zone the part of the ocean, near the surface, that receives enough light to sustain photosynthesis.

photosynthesis the process by which some organisms use the energy of sunlight to produce organic molecules, usually from carbon dioxide and water.

phytoplankton tiny photosynthetic organisms that float near the ocean surface.

plankton animals and plants that float passively in the ocean.

plankton bloom the sudden and rapid multiplication of plankton that results in dense concentrations of planktonic organisms.

plate tectonics the movement of large segments of the Earth's outer crust and mantle (lithospheric plates) relative to one another; the new paradigm of earth sciences developed through the 1960s and 1970s.

primary producer those organisms in a food chain, such as green plants and photosynthesizing or chemosynthesizing bacteria, upon which all other members of the food chain depend directly or indirectly; those organisms not dependent on an external source of nutrients; also known as autotrophs.

primary productivity the quantity of organic matter that is synthesized from inorganic materials by autotrophs.

prokaryote a unicellular organism lacking a nucleus and such organelles as plastids and mitochondria; cells of archea, bacteria, and cyanobacteria.

protist a member of the kingdom Protista, which contains any eukaryotic organism that is not a plant, animal, or fungus.

protozoan belonging to the phylum protozoa; unicellular organisms that exhibit animal-like characteristics.

radiolarian a protozoan that has an intricate shell made of silica and that uses pseudopods to capture prey.

red tide a condition that occurs as the result of a population explosion of certain dinoflagellates (or other protists) that imparts a red colour to the water.

reservoir rock sedimentary strata having porous and permeable properties and acting as a reservoir for hydrocarbons.

rift valley the fault-bounded valley found along the crest of many ocean ridges, created by tensional stresses that accompany the process of seafloor spreading; also occurs on continents.

rogue wave an unusually large and dangerous wave, usually short-lived and occurring in the open ocean, that is created by constructive wave interference.

salinity a measure of the total concentration of dissolved solids in seawater, generally expressed in parts per thousand.

seafloor spreading the process by which oceanic crust is created at the crest of ocean ridges, and tectonic (lithospheric) plates diverge.

seamount a submarine mound, usually of volcanic origin, that rises sharply from the seafloor.

sediment grains or particles of either inorganic or organic origin deposited by air, water, or ice.

sediment wave a series of sediment layers arranged in large-scale, regular undulations, with a typical wavelength (crest to crest) of around 1 kilometre; formed by deposition of fine muddy sediment from turbidity currents and bottom currents.

sedimentary basin thick accumulation (typically 1–15 kilometres thick) of sediments in a particular region, occurring below both the continents and seafloor.

sedimentary cycle the process by which sediments are formed, transported, deposited, and then compacted and cemented into sedimentary rocks, before being uplifted and once more subjected to weathering and erosion.

sedimentary rock a rock formed from the compaction and cementation of sediment; one of the three principal rock types.

seismic pertaining to a naturally occurring or artificial earthquake or earth vibration.

seismic surveying the use of sonar and explosive devices and measuring equipment to study the nature of earthquakes and the layered structure of the Earth's crust, mantle, and core.

shelf sea the area of ocean found at the edge of the continental shelf, before the continental slope.

siliceous material whose composition is silica (SiO_2); generally referring to biogenically produced siliceous material (from diatoms, radiolarians, etc.).

spreading centre mid-oceanic ridges where seafloor spreading occurs.

stratigraphy the branch of geology that studies the age relationship and significance of layered sedimentary rocks and the sequence of fossils they contain; also known as historical geology.

stratosphere the lowest part of the middle atmosphere, 12–45 kilometres above the Earth's surface.

stromatolite laminated calcareous sedimentary formation produced by lime-secreting cyanobacteria.

strike-slip margin a boundary at which tectonic plates move past one another in a lateral direction, associated with intense earthquake activity; also known as transform faults beneath the ocean.

subduction the movement of one lithospheric plate underneath another so that the descending plate is consumed into the mantle; generally creates a deep oceanic trench at the point of descent.

subduction zone an area where subduction is occurring.

submarine canyon deeply incised, steep-walled valley, commonly V-shaped in profile, which cuts into the rocks and sediments of the outer continental shelf and the continental slope.

submarine fan a cone-shaped sedimentary deposit that accumulates on the continental slope and rise; generally fed from a distinct point or line-source of sediment such as a major river or delta.

submarine waterfall major overflow of bottom-water mass between one ocean basin (or compartment) and another; the total drop may exceed 1 kilometre but the actual gradient is very low, more like a river rapids than a true waterfall.

subsidence the sinking of large portions of the Earth's crust under the influence of major tectonic forces.

substrate a general term in reference to the surface on or within which organisms live or onto which sediment is deposited.

symbiosis an intimate living relationship between two different organisms.

tectonic uplift a rise in topographic height due to crustal movement caused by major tectonic forces.

terrigenous sediment sediments derived from the weathering and erosion of pre-existing rocks.

Tethys Ocean a former ocean that separated Gondwanaland from Laurasia during the early Mesozoic Era.

thermocline a sharp, vertical temperature gradient that marks a transition zone between water masses having markedly different temperatures.

thermohaline the vertical movement of water masses that results from differences in density caused by differences in salinity and/or temperature.

tidal range the vertical difference separating the water level between successive high and low tides.

transform fault a steep boundary separating two lithospheric plates along which there is lateral slippage.

trench long, narrow, and deep topographic depression associated with a volcanic arc, that together mark a collision or subduction zone where one lithospheric plate is overriding another.

trophic level a functional or process category for types of feeding by organisms.

troposphere the lowest part of the atmosphere, extending to around 10–15 kilometres.

tsunami a destructive sea wave that is usually produced by an earthquake but can also be caused by submarine landslides or volcanic eruptions.

turbidite sediment layer, typically with graded bedding, which is deposited by a turbidity current.

turbidity current a density-driven current of sediment-laden water that flows swiftly downslope, in some cases traveling many hundreds of kilometres from the shelf edge to the abyssal plain.

upwelling the slow upward transport of water to the surface from depth; generally recycling nutrient elements and organic material.

volcanism a number of processes associated with the release of magma at the Earth's surface, generally accompanied by hot water, steam, and other gases.

water mass a body of water identifiable from its temperature, salinity, or density.

wave interference the meeting of two or more separate wave trains, leading to an increase and/or decrease in wave height, or to a complex cross-cutting wave pattern.

wave train a series of waves coming from the same direction and having a distinctive pattern of wave height and period.

zooplankton animal plankton such as foraminifera and radiolaria.

zooxanthellae symbiotic dinoflagellates that live in the tissue of corals and other reef-building organisms.

Further reading

Beerling, D., 2007, *The Emerald Planet*, Oxford University Press.

Benton, M.J., 2003, *When Life Nearly Died: The Greatest Mass Extinction of All Time*, Thames and Hudson.

Bigg, G.R., 1996, *The Oceans and Climate*, Cambridge University Press.

Byatt, A., Fothergill, A., and Holmes, M., 2002, *The Blue Planet: A Natural History of the Oceans*, BBC/DK.

Deacon, M., Rice, T., and Summerhayes, C., 2001, *Understanding the Oceans*, UCL Press.

Dixon, D., Jenkins, I., Moody, R., and Zhuravlev, A., 2001, *Cassell's Atlas of Evolution*, Cassell & Co.

Ellis, R., 2000, *Encyclopedia of the Sea*, Knopf.

Fortey, R., 1999, *Life: A Natural History of the First Four Billion Years of Life on Earth*, Vintage.

Frances, P. and Guerrero, A.G. (eds), 2006, *Ocean: The World's Last Wilderness Revealed*, Dorling Kindersley.

Gage, J. and Tyler, P., 1992, *Deep-sea Biology*, Cambridge University Press.

Holland, H.D. and Petersen, U., 1995, *Living Dangerously: The Earth, Its Resources and the Environment*, Prentice Hall.

Karleskint, G., 1998, *Introduction to Marine Biology*, Harcourt Brace & Co.

Kunzig, R., 2000, *Mapping the Deep: The Extraordinary Story of Ocean Science*, Sort of Books.

Jones, S., 2001, *Almost Like a Whale*, Black Swan.

Levinton, S., 2010, *Marine Biology: Function, Biodiversity, Ecology*, Oxford University Press.

Marshak, S., 2005, *Earth: Portrait of a Planet* (2nd edition), W.W. Norton & Co.

Middleton, N., 1999, *The Global Casino: An Introduction to Environmental Issues*, Arnold Publishers.

Mladenov, P.V., 2013, *Marine Biology: A Very Short Introduction*, Oxford University Press.

Monahan, D., 2001, *World Atlas of the Ocean*, Firefly.

Monroe, J.S. and Wicander, R., 2001, *The Changing Earth: Exploring Geology and Evolution*, Brooks/Cole.

Pickering, K.T. and Owen, L.A., 1997, *An Introduction to Global Environmental Issues* (2nd edition), Routledge.

Pinet, P.R., 2006, *Invitation to Oceanography* (4th edition), Jones & Bartlett Publishers.

Press, F. and Siever, R., 2001, *Understanding Earth* (3rd edition), W.H. Freeman.

Redfern, R., 2000, *Origins: The Evolution of Continents, Oceans and Life*, Cassell & Co.

Roberts, C., 2007, *Unnatural History of the Sea*, Island Press.

Roberts, C., 2013, *Ocean of Life*, Penguin Books.

Southwood, R., 2003, *The Story of Life*, Oxford University Press.

Stanley, S.M., 1989, *Earth and Life Through Time* (2nd edition), W.H. Freeman.

Stow, D.A.V., 2004, *An Encyclopedia of the Oceans*, Oxford University Press.

Stow, D.A.V., 2005, *Oceans: An Illustrated Reference*, University of Chicago Press.

Stow, D.A.V., 2010, *Vanished Ocean: How Tethys Reshaped the World*, Oxford University Press.

Summerhayes, C.P. and Thorpe, S.A., 1996, *Oceanography: An Illustrated Guide*, Manson Publishing.

Thurman, H.V. and Trujillo, A.P., 1999, *Essentials of Oceanography* (6th edition), Prentice Hall.

Van Andel, T.H., 1994, *New Views on an Old Planet: A History of Global Change* (2nd edition), Cambridge University Press.

Walker, G. and King, D., 2008, *The Hot Topic: How to Tackle Global Warming and Still Keep the Lights On*, Bloomsbury.

Woodward, C., 2001, *Ocean's End: Travels Through Endangered Seas*, Basic Books.

Index

C

Oceans

Oceans

Expand your collection of
VERY SHORT INTRODUCTIONS